GOVERT SCHILLING

GALAXIES

A Firefly Book

Published in paperback by Firefly Books Ltd. 2023
Text © 2018 Govert Schilling
Photographs © as listed on page 234
English edition © 2019 Firefly Books Ltd.
© 2018 Fontaine Uitgevers
Original title: Schilling, Galaxies

All rights reserved. No part of this publication may be reproduced, stored in a retrieval system, or transmitted in any form or by any means, electronic, mechanical, photocopying, recording or otherwise, without the prior written permission of the Publisher.

First printing

Library of Congress Control Number: 2023932943

Library and Archives Canada Cataloguing in Publication
Title: Galaxies : birth and destiny of our universe /
    Govert Schilling.
Other titles: Galaxies. English
Names: Schilling, Govert, author.
Description: Includes index. | Translation of:
    Galaxies: sterren als bouwstenen van het heelal.
Identifiers: Canadiana 20230177476 |
    ISBN 9780228104483 (softcover)
Subjects: LCSH: Galaxies—Popular works. |
    LCSH: Galaxies—Pictorial works. | LCSH: Galaxies.
Classification: LCC QB857 .S3513 2023 |
    DDC 523.1/12—dc23

Published in Canada by
Firefly Books Ltd.
50 Staples Avenue, Unit 1
Richmond Hill, Ontario
L4B 0A7

Published in the United States by
Firefly Books (U.S.) Inc.
P.O. Box 1338, Ellicott Station
Buffalo, New York
14205

Translation: Travod International Ltd.

Printed in China

GOVERT SCHILLING

GALAXIES

*Birth and destiny
of our universe*

FIREFLY BOOKS

CONTENTS

Introduction 6

1. Our Milky Way 16
- Cosmic Delivery Rooms 18
- Stars and Planets 26
- When Stars Die 34
- The Center of the Milky Way 42

Intermission
- Surveying the Milky Way 50

2. Cosmic Neighbors 52
- The Magellanic Clouds 54
- The Andromeda Galaxy 62
- The Triangulum Galaxy 70
- Satellite Galaxies 78

Intermission
- How Far Away Is this Star? 86

3. A Gallery of Galaxies 88
- Spiral Galaxies 90
- Barred Spiral Galaxies 98
- Ellipses, Lenses and Dwarfs 106
- Dark Matter 114

Intermission
- The Expanding Universe 122

124 Monsters and Gluttons
126 Dancing Galaxies
134 Collisions and Mergers
142 Active Cores and Quasars
150 Supermassive Black Holes

Intermission
158 Big Eyes

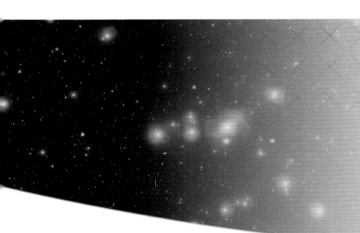

160 Galaxy Clusters
162 Cosmic Collections
170 Gravitational Lenses
178 Dark Forces
186 The Structure of the Universe

Intermission
194 A Glance into the Past

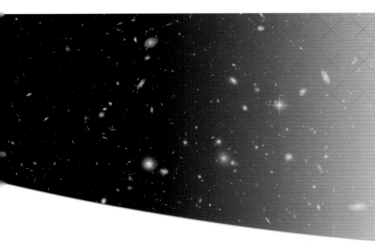

196 Birth and Evolution
198 At the Edge of Space
206 The First Galaxies
214 The Dawn of the Universe
222 Dark Energy

Intermission
230 Precision Cosmology

234 Picture Credits
236 Index

Introduction

Grab a pin with a colored head from your pincushion or bulletin board. One of those pins with the colored ball at the top. On a starry night, take it outside and, with an outstretched arm, point the sharp end toward the sky. Now close one eye and look at it. That minuscule spot of the starry sky covered by the pinhead contains several thousand galaxies an unimaginable distance away. Each of these spots is comparable to our own Milky Way. They're all gigantic collections of tens or hundreds of billions of stars.

Theologians once argued about how many angels can dance on a pinhead. Astronomers take a different approach: they calculate how many galaxies disappear behind a pinhead.

The observable universe contains at least a hundred billion such systems. Collectively, these galaxies easily contain about ten trillion stars like our Sun – about a hundred times as many as the total amount of all the grains of sand found in all the deserts on Earth. If stars are the inhabitants of the cosmos, then galaxies are the villages and cities – starting with small, structureless dwarf galaxies up to gigantic spiral systems, comparable to Earthly metropolises. And just as human agglomerations aren't arbitrarily scattered over the Earth's surface, the villages and cities in the universe are also grouped into clusters and superclusters, the largest contiguous structures in the three-dimensional cosmic landscape.

Galaxies are essentially the building blocks of the universe. However, no matter how large, important and numerous they may be, no one knew of their existence until a century ago. Although small, hazy flecks of light were discovered on the firmament, their true nature was unknown; many astronomers thought they were emerging stars – rotating gas nebulae in our own Milky Way. Incidentally, it wasn't until the middle of the 20th century that the spiral structure of the Milky Way was discovered.

For thousands of years, humans have been at the level of a cosmic kindergarten child, completely unaware of the wide world outside our immediate surroundings. Only in the last 100 years have we looked at what lies beyond the fence.

Distant neighbors

To mark its 27th birthday, the Hubble Space Telescope took this image of galaxies NGC 4298 and NGC 4302, both 55 million light-years from Earth, in the constellation Coma Berenices, in spring 2017. We see one galaxy diagonally from above, its spiral structure clearly visible, and the other one is seen more or less from the side, highlighting its dark dust clouds.

Astronomers deciphered the structure of our own star city, befriended the other inhabitants and recorded the biographies of distant suns. During the last century, it was finally recognized that there was more than just our own Milky Way and that the universe was, in fact, an implausible extension of space and time filled with galaxies of various shapes and forms.

Thanks to the Hubble Space Telescope's sharp vision and high sensitivity, many galaxies have become visible down to the last detail. Graceful disks, gravitational spirals, symmetrical halos – each galaxy is a unique feast for the eyes. Astronomers have also discovered an equally diverse range of star clusters and nebulae, exploded stars and mysterious halos, colliding galaxies and huge black holes.

No less fascinating is the fact that distant galaxies give us a glimpse into the earliest days of the universe. The light from these distant star clouds has traveled to us for billions of years, allowing us to look back into the dawn of the very first galaxies. They were almost formless, barely illuminated dots and streaks – nothing more. And yet these dots and streaks show us what the cosmos looked like billions of years ago, long before the emergence of the Sun and Earth.

This book will take you on a journey from our own familiar Milky Way to the farthest boundaries of time and space. With over 160 carefully selected photos and illustrations, I present the latest scientific findings on galaxies, quasars, galaxy clusters, gravitational lenses and the history of the universe.

If you want to explore the structure and evolution of the universe, you have to look at the fascinating world of galaxies. For this book, I was gratefully able to draw from the impressive photo collections of the European Southern Observatory and the Hubble Space Telescope. This book is thus also an ode to the persistence of the engineers who developed the telescopes and instruments with which we explore the universe and to the astronomers who generously share their research with the rest of the world.

I hope that reading – and observing! – this book will astonish the reader as much as it has astonished me while writing and compiling it.

Govert Schilling

A colorful world of stars

Galaxies are the villages and cities of the universe. They are where stars live out their lives, from their birth in multicolored nebulae of gas and dust – such as the Lagoon Nebula, captured in this Hubble photo – to their death in catastrophic supernova eruptions.

Doppelgänger

The magnificent spiral galaxy NGC 6744 is located 30 million light-years from Earth, in the constellation Pavo. Our own Milky Way would look quite similar if viewed from that distance. The reddish spots that can be seen in the spiral arms are concentrations of luminous hydrogen gas, where new stars are born. This photo was taken with the 7¼-foot (2.2 m) telescope of the European Southern Observatory in La Silla, Chile.

A nearby colossus

NGC 5128 is a huge elliptical galaxy situated just 12 million light-years from Earth. In this photo, the glowing embers of many billions of stars are partially hidden from view by a broad, curved band of dark dust. Thousands of globular clusters have been discovered around NGC 5128. This galaxy,- also known as Centaurus A – contains a gigantic black hole at its center, and it is also a powerful radio source.

Double ring

The NGC 7098 galaxy appears to have a double ring structure. It is, in fact, an extraordinary barred spiral galaxy; the outermost "ring" of stars actually consists of two spiral arms. This galaxy is located 95 million light-years from Earth, in the southern constellation Octans. In the background of the photo you can see countless small, very distant galaxies.

An exceptional view

From one horizon to the other, the band of the Milky Way spans the four gigantic buildings that make up the European Southern Observatory's Very Large Telescope, on top of the Cerro Paranal mountain in Northern Chile. Bizarre dust clouds obstruct the view of the bright center of the galaxy. The pink spots are active regions of star formation. From our earthly perspective, the Milky Way provides the "interior view" of a galaxy.

Our Milky Way

Cosmic Delivery Rooms

Orion's secrets

A long exposure photograph of the famous winter constellation Orion shows the contours of the Orion Molecular Cloud Complex, an active star-forming region in the Orion Nebula, which is a bright cloud south of the constellation's three belt stars. This cosmic birthplace is about 1,350 light-years away from us. The orange star in the upper left corner is Betelgeuse.

Rarely have I seen the Milky Way as beautiful as I did in the spring of 1998. The construction of the European Southern Observatory's (ESO) Very Large Telescope was in full swing on top of Cerro Paranal, a 8,530-foot (2,600 m) high mountain peak located in inhospitable northern Chile. Greek astronomer Jason Spyromilio, who later became director of the observatory, showed me around during the day. In the evening, we had dinner in the canteen at the base camp – a large collection of sea containers that had been converted into simple sleeping quarters. The area was illuminated by spotlights. Suddenly the power went out.

It took about half an hour for them to get an emergency generator started. During this time, everyone came outside – construction workers, engineers and astronomers – to see an impressive starry sky most had never seen before: a velvet-black firmament covered with thousands of glittering spots of light. Directly above our heads was the wide band of the Milky Way – an interior view of our own galaxy.

We watched 400 billion stars circling in a slow procession around the center of a gigantic flattened disk, like pilgrims circling the Kaaba in the trance-like ceremony during the Hajj. In the Great Mosque of Mecca, a pilgrim can see his fellow worshippers in every horizontal direction around him, but none are above or below him. Similarly, from our perspective here on Earth, we see the stars melting into a band of lights spanning the celestial dome on the outskirts of the Milky Way. It's a patchy band interspersed with star clouds and nebulae, where frayed dust clouds block our view of the bright center. The glow of millions of stars many thousand light-years away – a more impressive picture of our own nothingness is hardly possible. (A light-year is the distance that light travels at a speed of 186,000 miles per second [300,000 km/s] in a year; it corresponds to about 5.9 trillion miles [9.5 trillion km].)

A look at the Milky Way is a humbling realization of humanity's place in space and time. For billions of years, this vast spiral system has carried out the cosmic cycle of the birth, life and death of stars. Only 4.6 billion years ago, when the Milky Way had already reached two-thirds of its present age, our Sun emerged as an inconspicuous dwarf star in the midst of countless inconspicuous peers.

Debris and dust – remnants of this birth – clumped together to form planets. On one of these planets, organic molecules developed into self-aware beings marveling at the universe from a pitch-black mountaintop.

For thousands of years, humans have filled the celestial sphere with gods, mythical creatures and other fantasy figures. The Milky Way was considered a heavenly river or the road to the afterlife. It was only a few hundred years ago – a cosmic blink – that mythology gave way to science, although facts can often seem stranger than fantasies. Who could have guessed that the atoms within our own bodies were forged within other suns?

Apart from a random meteor or a rare comet, the starry sky appears eternal and unchangeable, the Milky Way being the epitome of cosmic stability. But appearances can be deceiving. Above all, it's the transience of humanity that drives us to the idea of a heavenly stillness. A human life is nothing more than a mere breath in the life of a star. Since the birth of *Homo sapiens*, the Sun has completed only one thousandth of its orbit around the Milky Way.

Chaos in Carina

This mosaic, composed of 48 Hubble photographs, shows the Carina Nebula, in the southern constellation Carina (the ship's keel) – a chaotic complex of gas and dust clouds situated 7,500 light-years from Earth, in which new stars are continually being born. Shock waves and young protostars can be seen throughout the nebula. Strangely shaped dust clouds stand out darkly against the bright background.

A colorful delivery room

Like the Orion Nebula, NGC 2467 is a gigantic star-forming region located 13,000 light-years from Earth. In this photo, dark clouds of dust darken the colorful background of luminous hydrogen gas. Long streaks and "fingers" stretch up, similar to the dust columns in the Eagle Nebula (on the following page). Most of the stars in this photo are at most a few dozen million years old.

Eratosthenes and Einstein, Huygens and Hawking – every natural scientist from the past, present and future is actually researching the same still image of the Milky Way.

In order to accurately represent this galactic biography, we must examine periods of time spanning hundreds of millions of years. If we could compress eons into minutes, we would see cosmic clouds bubbling up and colliding, gas and dust nebulae collapsing and fragmenting, and newborn stars twinkling like fireflies. We'd experience how the process of star formation spreads like a sprawling forest fire and how the catastrophic death of one star causes the birth of another.

The Milky Way is the hectic kitchen of a hundred billion starry restaurants, where the laws of nature dictate the recipes and no chef is needed. Take plenty of hydrogen and helium, a pinch of heavy elements and let gravity do the rest. Before you know it, the first suns will be served up and, with a little luck, planets as well.

Everything begins with dark molecular clouds – huge, cold nebulae in which atoms have joined together to form simple molecules of hydrogen and carbon monoxide. With dimensions of hundreds of light-years, they're only visible when they emerge like a dark silhouette against the bright nebular background or when we capture their microwave radiation using special parabolic antennas, like those of the ALMA Observatory in northern Chile.

Where the density of the gas is highest, gravity triumphs. Atoms and molecules then become more and more attracted to each other. Next, a compact nucleus forms in the center of the cloud, and this nucleus contains enough building materials to form dozens, or perhaps even hundreds, of stars. Shaken by turbulence and magnetic fields, the nucleus decays into countless fragments – these are the embryos of single stars, such as the Sun, and double stars and triple or quadruple star systems.

After a short period of at most a few hundred thousand years, a completely new star cluster shines deep within the molecular cloud. Once their atomic furnaces are lit, the newborn stars blow energy-rich radiation out into space. The cloud enveloping them is heated from the inside and then blown away. New

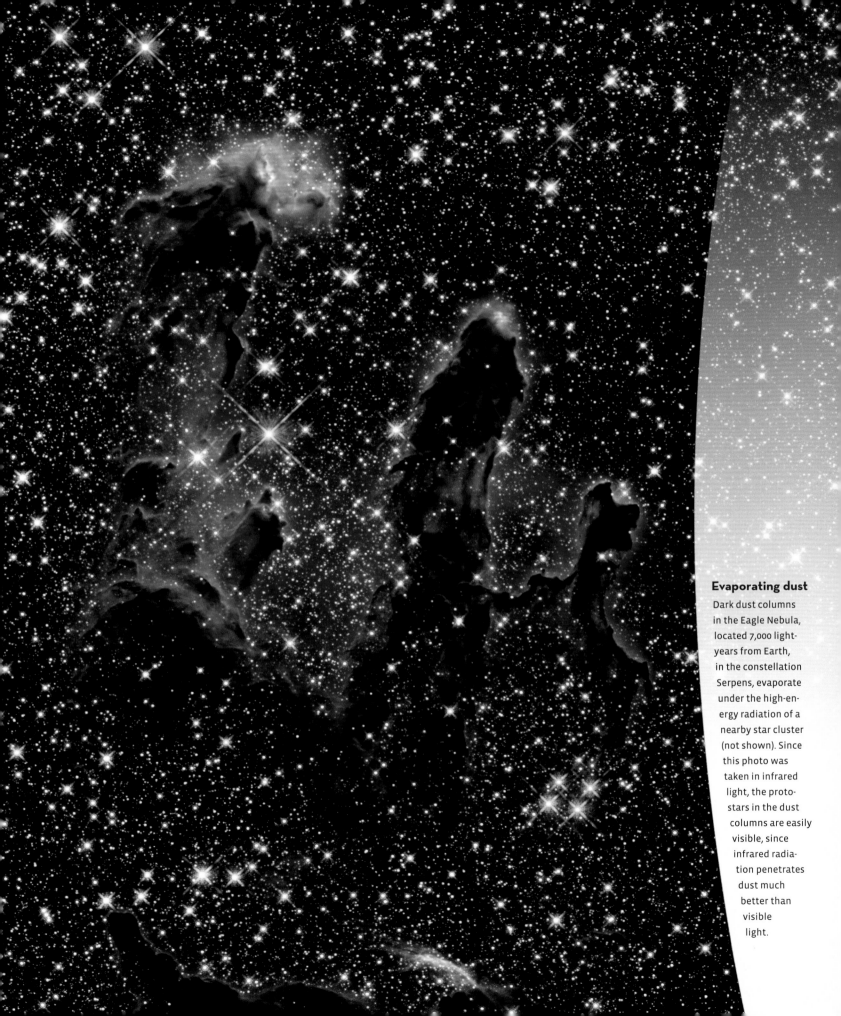

Evaporating dust

Dark dust columns in the Eagle Nebula, located 7,000 light-years from Earth, in the constellation Serpens, evaporate under the high-energy radiation of a nearby star cluster (not shown). Since this photo was taken in infrared light, the protostars in the dust columns are easily visible, since infrared radiation penetrates dust much better than visible light.

stars emerge from the shock waves created in the gas and dust. And so one newborn cry after another resounds in the cosmic delivery room. The dark cloud – the uterus of the star-forming process – slowly dissolves. The ultraviolet radiation of the heaviest newborns in the central star cluster penetrates the volatile gas and shatters the molecules into unbound atoms, turning them into a peculiar glowing rosy flame.

Elsewhere, bizarre gas and dust clouds are compressed by the same radiation and transformed into new star embryos – cosmic "eggs" that will hatch in the not too distant future. At the same time, the rays gnaw away at the edges of these dust clouds. Just as sandstone formations erode over tens of thousands of years under the effects of the wind, so too are the dark clouds gradually eroded by the effects of starlight. Only in the shade of the largest areas of compression, on the leeward side of the wind, does the dust persist. In this way, elongated dark fingers emerge, seemingly pointing at the central star cluster, until they also bare new stars and then completely evaporate. This fate also awaits young star clusters.

Within a hundred million years – just one percent of the age of the Milky Way – the sparkling accumulation gradually falls apart as gravity no longer binds the stars together, so they roam in every direction. The heaviest family members have already ended their short lives, while lighter brothers and sisters, like our own Sun, go out into the wide world, traveling along the banks of the Milky Way.

From Cerro Paranal, the Milky Way offers a breathtaking sight: the glow of millions of distant stars that can only be seen separately with a powerful telescope.

Elsewhere in the sky, there are thousands of stars in our immediate vicinity. Have the nearest relatives of our own Sun mixed in with them? Stars whose light of life ignited a good four and a half billion years ago in the same cluster of stars? No one knows. But we do know that at this very moment new stars are being born in countless places in the Milky Way, hidden in the caverns of the gas and dust clouds, which stand out as dark contrasts against the brightly shining band of the Milky Way.

There will be new suns, new planets, maybe even new life. This miracle occurs everywhere in the Milky Way. It began long before the Earth came into being and will continue long after humans disappear. As a cosmic ephemera on a wandering grain of sand, I can't get enough of this spectacle.

One big family

NGC 6611 is a 5.5 million-year-old star cluster. The stars were formed around the same time as those found in the center of the Eagle Nebula. In this photo, we can see that the newborn stars have blown all of the dust and debris out of the nebula, giving it a "clean" appearance. In about a hundred million years, the star cluster will break apart, and the stars will spread throughout the Milky Way.

Stars and Planets

The Milky Way is a cosmic metropolis populated by several hundred billion inhabitants. Complacent couples, idiosyncratic singles, homey families, spitfires, eccentric creatures, it's a colorful melting pot of young and old, hot and cold, big and small, heavy and light, rich and poor. No two stars are alike; each has its own distinctive fingerprint, which can be read using sensitive spectroscopes (which measure the light they emit). But the stars do all have one thing in common: they are all gigantic balls of glowing hot gas and are driven by the nuclear fusion reactions inside them.

Nuclear fusion sounds more complicated than it actually is. Just as two small companies occasionally merge into a larger one, two light atomic nuclei can merge into a heavier one. During this process, some mass is converted into energy, as Albert Einstein's famous formula $E = mc^2$ puts it. It is this energy that is

A teenage star with jets

Behind the opaque dust clouds in this Hubble photo of the Orion Nebula, there is a young protostar. What is visible are two stellar jets being blown into space from the protostar's north and south poles. These narrow bundles of fast-moving hot gas dissipate a great deal of rotational energy. At the point where the gas collides with the surrounding matter, shock waves, which can be seen as small, bright regions, are generated.

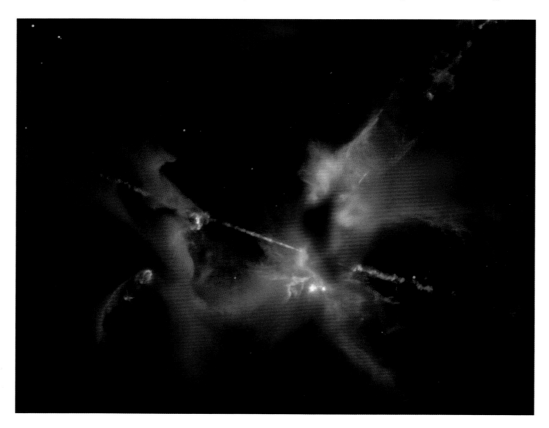

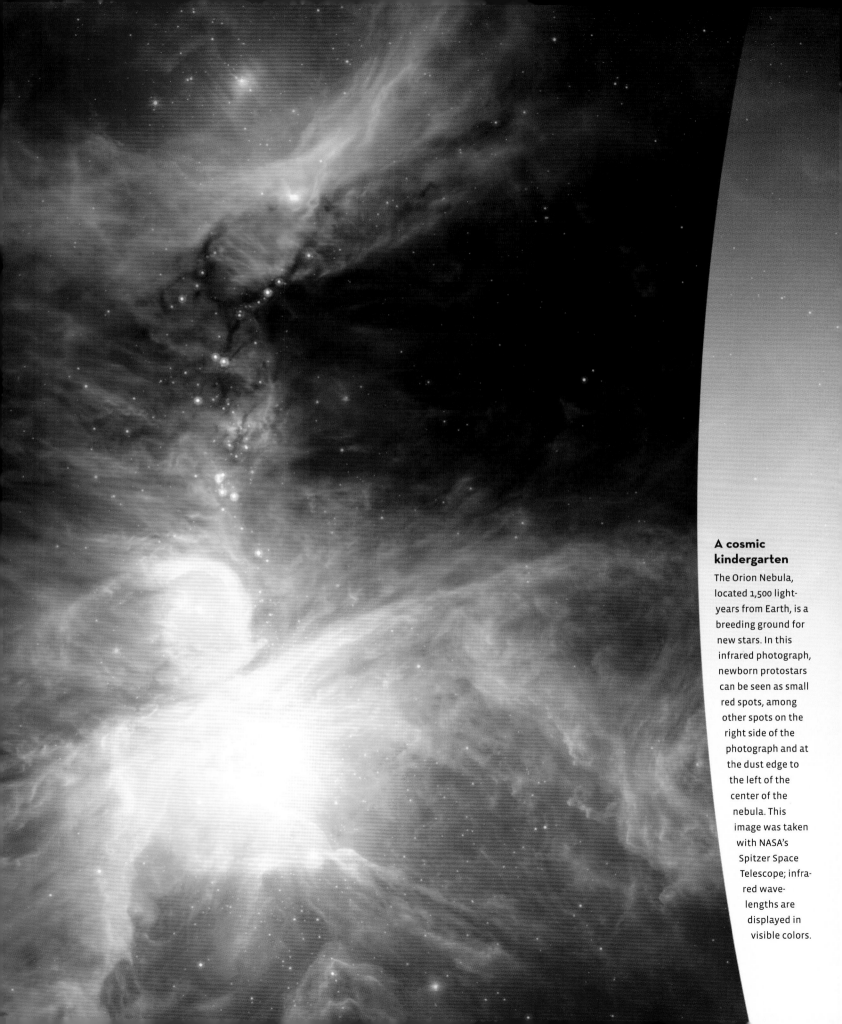

A cosmic kindergarten

The Orion Nebula, located 1,500 light-years from Earth, is a breeding ground for new stars. In this infrared photograph, newborn protostars can be seen as small red spots, among other spots on the right side of the photograph and at the dust edge to the left of the center of the nebula. This image was taken with NASA's Spitzer Space Telescope; infrared wavelengths are displayed in visible colors.

Worlds in the making

Dust and rock particles in a protoplanetary disk clump together relatively quickly, forming larger chunks, boulders and finally complete planets. This image is based on astronomical observations; it shows that the interior of the disk has been blown clean by the energetic radiation of the young star at the center.

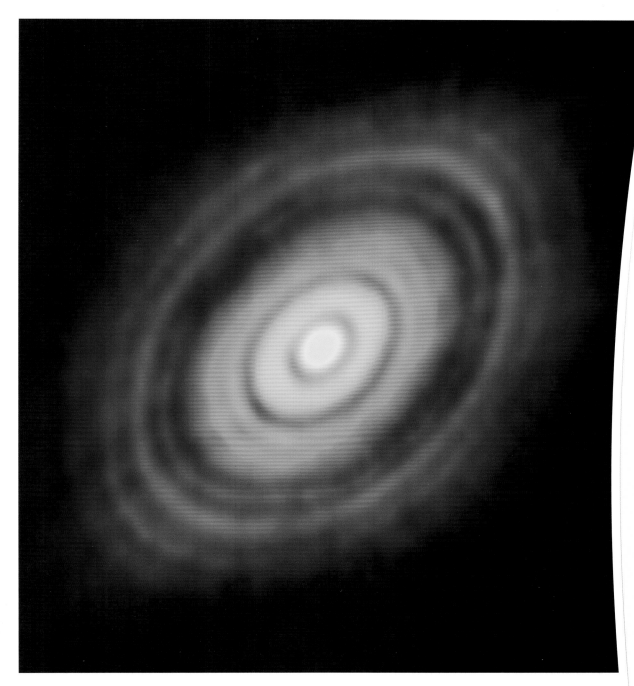

Disks and rings
The ESO's ALMA Observatory has recorded the extended gas and dust disk around the protostar HL Tauri with unprecedented detail. The dark zones in the disk are probably caused by gravitational disturbances of newborn planets. HL Tauri is less than a million years old and is located about 450 light-years from Earth.

ultimately emitted by the surface of a star in the form of light and other radiation. Each star in the night sky is a cosmic nuclear power plant.

Under earthly conditions, nuclear fusion is a difficult undertaking, since the atomic nuclei must be greatly compressed. Although we "succeeded" in building a hydrogen bomb, controlled nuclear fusion for peaceful energy generation is still a dream of the future. Inside a star, on the other hand, there is by nature tremendous pressure and a correspondingly high temperature. Atomic nuclei lie shoulder to shoulder, so to speak, and nuclear fusion reactions occur spontaneously and continuously under these extreme conditions.

However, this process does require a minimum quantity of gas. If an interstellar shrinking gas cloud is less than 14 times heavier than Jupiter, then a gaseous celestial body in the shape of a perfect sphere is formed, and its internal pressure and temperature are not high enough to trigger nuclear fusion reactions. Perhaps billions of such cool, dark balls of gas roam throughout the Milky Way like lonely wandering giant planets without a mother star. However, if the shrinking gas cloud weighs more, the pressure and temperature rise sufficiently to fuse deuterium atoms.

But deuterium – heavy hydrogen – doesn't occur often in the universe, so not much energy is produced. The result is a gently glowing star, hardly bigger than a planet, that emits little heat and almost no visible light. Such "brown dwarfs" are hard to spot yet are probably quite common.

Only when a ball of gas exceeds the weight of Jupiter by more than 70 times can we speak of a real star in which protons – the nuclei of hydrogen atoms – fuse to form heavier helium atomic nuclei. The more gas is available, the higher the internal pressure and temperature will be, and the more efficient the fusion process will be. The initial mass thus determines the appearance and character of the resulting star, from small, cool dwarf stars to large, hot giant stars. The latter burn their available fuel at such an incredibly high rate that they reach the end of their life within a few million years. In contrast, cool red dwarfs consume very little hydrogen and can live for many billions of years.

How exactly a large, volatile cloud of interstellar gas collapses under its own weight and forms a shining star is still not fully understood. Scientists theorize that gravitational force pulls the gas particles toward each other, but as the cloud becomes smaller and more compact, it begins to rotate faster. Centrifugal forces transform the cloud into a flat rotating disk. In order for a smooth rotating star to form in the center, this gas and dust disk must lose much of its rotational energy. This likely occurs by means of powerful stellar jets – whirling "jet streams" of gas that are blown into space in opposite directions along the axis of rotation of the young protostar. But how exactly these jets are created is unclear.

A slowly rotating protostar remains and is surrounded by a disk of gas and dust in which planets can clump together within a few hundred thousand years. It's a strange thought: a complete world like Earth, with its oceans, deserts and volcanoes, is nothing more than a waste product of the birth of a single star. All planets, moons, comets and planetoids together make up only one percent of the total mass of the solar system; the remaining 99 percent is contained within the Sun. Equally inconceivable is the fact that almost all stars in space are accompanied by a planetary system and that the number of Earth-like planets in our Milky Way amounts to tens of billions.

Protoplanetary disks around young stars have been explored in detail. Empty zones in such a disk reveal the presence of newborn planets. Spectrographs with high sensitivity on large earthbound telescopes and space telescopes like Kepler have also discovered thousands of exoplanets around older stars, including stars like our Sun. We know their mass, their size and their composition; in some cases they're very similar to Earth. Nothing in the cosmos is unique, and our home planet is no exception. Whether there is water and life on these distant planets is still unknown. But even the nearest neighbors of the Sun – small, cool dwarf stars like Proxima Centauri and Trappist-1 – have revealed planets in the habitable zone of the mother star, which is, just far enough from it for liquid water to be present.

On a clear, dark night, when you look at the magical band of the Milky Way, remember that these billions of stars are accompanied by planets and that the building blocks of life – hydrocarbons and amino acids – are present everywhere in the cosmos. It's very unlikely that life formed only once in a single place in the universe. At the same time, we must understand that life here on Earth completely depends on the energy of the Sun – and the nuclear fusion reactions inside it. Cell division, reproduction, photosynthesis, evolution, consciousness – without billions of years of hydrogen to helium conversion, none of this would have happened.

We're completely dependent on the light and warmth of our star, this one luminous pinhole in the immense Milky Way. We are, therefore, also at the mercy of the unexpected tempers of the Sun: minute fluctuations in solar activity – still not properly understood, let alone predictable – have an impact on the Earth's climate. Occasional outbursts of lethal X-rays and electrically charged particles hit our planet and its delicate inhabitants.

Even if we can stand up to ice ages and heat waves, life on Earth will one day end. The luminosity of the Sun will gradually increase over the next hundreds of millions of years, causing the oceans to evaporate and turning the Earth into a dry, cooked hellscape. Nothing in the cosmos is unique and nothing lasts forever. The Sun is still a quiet representative of its species; our planetary system is a prime example of cosmic regularity, and Earth is a fertile oasis with exuberant life.

Elsewhere in the Milky Way, completely different scenarios are taking place. Stars swell to scorching red giants or are sucked up by black holes. Planets are thrown into space by gravitational disturbances or make an apocalyptic plunge into the bubbling outer layers of their mother star. There are collisions and explosions that truly go beyond our imagination.

Seven planets

Seven planets orbit around Trappist-1, a cool dwarf star situated 40 light-years from Earth. Two of these planets are in the "habitable zone" of the star. This illustration shows the view of the entire planetary system from one of the seven planets. Another planet stands in front of the star, creating a dark shadow.

Echoes of an explosion

At the beginning of 2002, the red giant star V838 Monocerotis (20,000 light-years from Earth) suddenly flared up, becoming a million times brighter than the Sun. For a brief period of time, it may have been the brightest star in the Milky Way. Years later, the Hubble Space Telescope photographed so-called light echoes around the star: the light of the eruption is reflected by a molecular cloud surrounding the star.

Senior stars

The spherical star cluster M 92, which lies 25,000 light-years from Earth, in the constellation Hercules, is a gigantic cluster of about 300,000 old stars. Spherical star clusters are among the oldest structures of the Milky Way; M 92 is roughly 12 billion years old. Compared to the stars in this globular cluster, the Sun is a newcomer to the cosmic stage.

Our existence may not be as self-evident as it seems. All the more fascinating is the realization that we're inextricably linked to this vibrant metropolis and all its galactic forces. *Homo sapiens* – evolving from primates to rational, self-confident beings – can't possibly be considered detached from the billions of years of cosmic evolution. We are literally stardust.

But romantic as our galactic origins may seem, they also have a dark side. We owe our creation and existence to the catastrophic end of many other stars, which, long before we ever existed, lived through the cycle of creation and decay. Life and death are inseparably connected in the universe. The time has come to visit the cosmic tombs of the Milky Way.

When Stars Die

Our Milky Way is a place of both birth and death. On average, a new star is formed once a year somewhere in a cosmic womb. There are, however, a similar number of deaths. Births and deaths are more or less in balance. What's more, the remains of the burnt out suns are recycled to create new stars and planets. The Milky Way is the stage of a majestic cosmic cycle to which Earth and humans are inextricably linked.

For small, cool red dwarf stars, the end of life lies in an immeasurably distant future. Over dozens of billions of years, these fuel-efficient stars will emit their dull light. Many of these galactic youngsters are accompanied by earth-like worlds orbiting around them, close to the weak source of light and heat in the middle. On such planets with enough heat for liquid water, the miraculous evolution of life has all the time in the world. Heavier stars like our own Sun burn their nuclear fuel at higher speeds. They age faster and are depleted sooner, thus burning out quicker.

The burial shroud of a star

The Helix Nebula is one of the Earth's closest planetary nebulae, being only 700 light-years away. In this infrared photo, you can see that the center of the nebula is filled with warm dust particles. In a few billion years, our own Sun will also be covered in a similar expanding gas shell, when it swells to a red giant and blows its outer gas layers into space.

The soap bubble within the swan

The Soap Bubble Nebula in the constellation Cygnus wasn't discovered until 2008. It, too, is a planetary nebula emitted by a dying star. The term "planetary nebula" was coined by William Herschel, who made many discoveries at the end of the 18th century. Through his telescope, he found that the nebulae looked like the pale planet disk such as Uranus. However, these nebulae have nothing to do with planets at all.

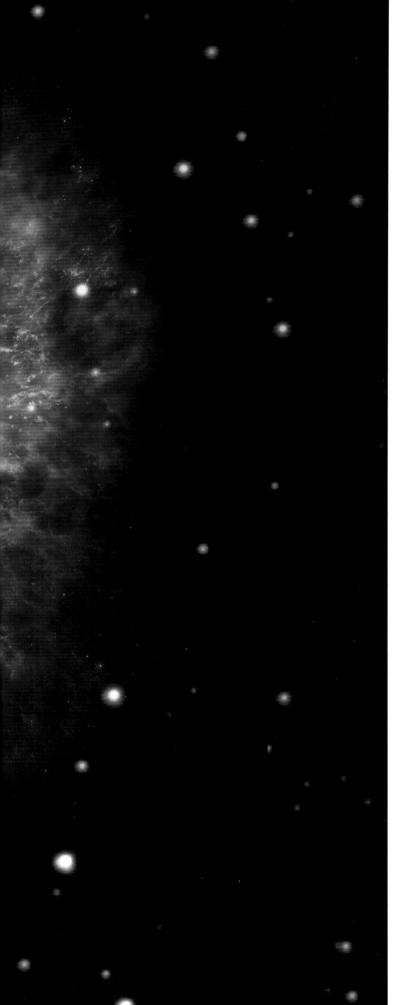

In the beginning, the temperature and luminosity of the Sun will gradually increase. In about a billion years, the oceans on Earth will slowly but surely evaporate; Mars, on the other hand, will begin to thaw. But this is only the beginning of the end. Only when the hydrogen supply inside the Sun is depleted and the helium atoms begin to fuse into carbon and oxygen, will the death sentence be sealed. Within a short time, the Sun will swell into a monstrous red giant star. On Mercury, the smallest, innermost planet, the temperature will rise so high that the rocky surface will become a glowing sea of viscous lava. Ultimately, the planet will be devoured by the ever-expanding red giant. Venus awaits the same fate: charred, vaporized and finally consumed. And when the dying star has reached almost 200 times its original circumference, not much more will remain of Earth than a scorched and dried-out lump of stone.

Around this time, the Sun will blast its outer gas layers into space and envelop itself for many thousands of years in a colorful, expanding nebula that will finally dissolve into the cosmos. Such "planetary nebulae" are scattered throughout the Milky Way – the last breaths of sun-like stars that were already fully grown when our Sun was born and are just now dying out. These include capricious layers of gas distorted by magnetic fields; complex formations like the Helix Nebula, with its radial tentacles like silent witnesses of a fierce death struggle; and delicate, symmetrical structures like the Bubble Nebula – all the result of a short, violent death. Eventually, nothing will remain of the dying star but a shriveled white dwarf that is hotter than the Sun but hardly bigger than Earth. A compressed ball of carbon and oxygen atoms with extremely thin outer layers of helium and hydrogen, it will fall victim to its own gravitational force, no longer able to resist nuclear fusion and radiation pressure.

Over the course of billions of years, the white dwarf star will cool down and become a cold, dark slag. The Sun's demise will follow this same pattern, and our once paradisiacal home planet – by then only a dark, sterile crumb – will be silent forever. Thus, the death of stars seems to offer little advantage to life in the cosmos. Yet one cannot exist without the other, for dying stars produce the chemical building blocks of plants, animals and humans. In the ejected gas layers of the dying red giants, atoms of hydrogen, carbon, oxygen and nitrogen string together to form simple organic molecules – the first step toward the formation of sugars, amino acids and DNA. To an even greater extent, catastrophic supernova explosions produce a similar result.

A colorful crab

The Crab Nebula is the remnant of a supernova explosion that lit up in 1054. At the center of this expanding nebula is a rapidly rotating neutron star, which can be seen from Earth as a so-called pulsar. This color photo was composed of images of several wavelength ranges: radio waves, infrared, visible and ultraviolet light as well as X-ray radiation.

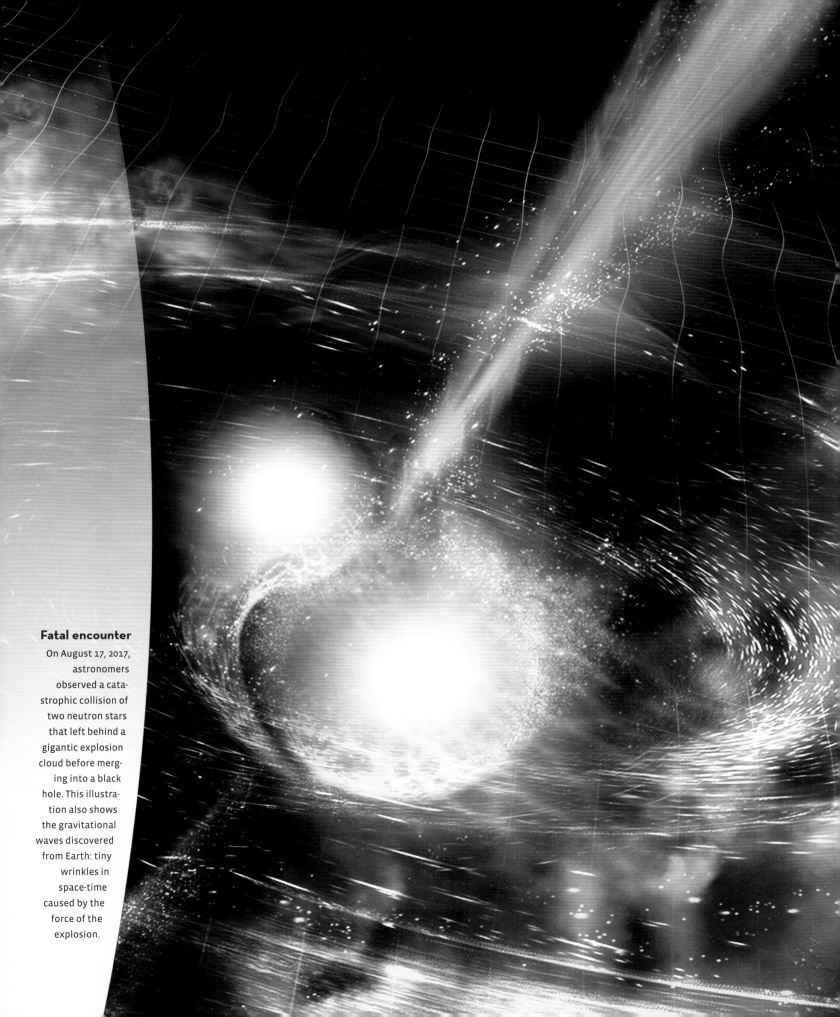

Fatal encounter
On August 17, 2017, astronomers observed a catastrophic collision of two neutron stars that left behind a gigantic explosion cloud before merging into a black hole. This illustration also shows the gravitational waves discovered from Earth: tiny wrinkles in space-time caused by the force of the explosion.

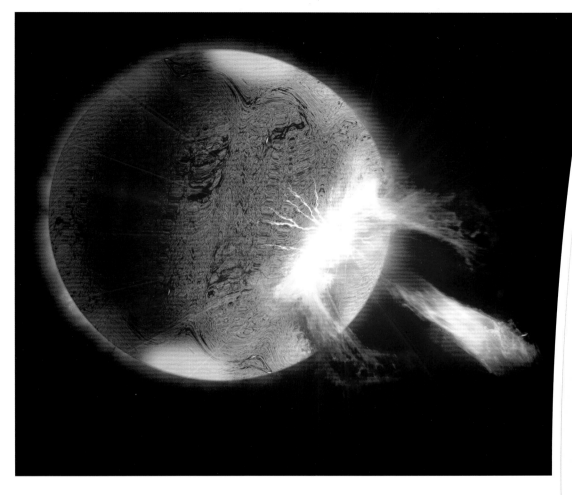

Magnetic monster

A neutron star is the ultra-compact remnant of a heavy star that ended its short life in a spectacular supernova explosion. Strong magnetic eruptions occasionally occur on the surface of this small, extremely compact and quickly rotating star, releasing as much energy in one millisecond as the Sun generates in 24 hours.

Without the nuclear fusion in the hot center of stars and without the formation of new elements in these celestial nuclear furnaces, the universe would still consist only of hydrogen and helium. Thanks to red giants losing mass and the heaviest stars exploding, newly formed elements have joined the cosmic cycle and a great chemical diversity has emerged.

The cradle of life stands in the cemetery of earlier generations of stars. Life on Earth owes its origin to one billion years of cosmic evolution. We are an integral and indispensable part of this evolving universe – every iron atom in our blood, every calcium atom in our bones and every carbon atom in our heart muscle was forged long ago in the nuclear fire of another star that once shone somewhere in the Milky Way. Without stellar winds, planetary nebulae and supernova eruptions, these atoms would still be trapped within other suns. We owe our existence to the mortality of the stars.

A supernova explosion defies description: a heavy star, about 20 times heavier than the Sun and of course much hotter and brighter, consumes its hydrogen supply in no time at all. Even the subsequent red giant phase, in which helium is burned to produce carbon and oxygen, is relatively short. But because of the extreme pressure and high temperature inside these stars, new fusion reactions begin that would never take place in lighter stars like the Sun. This in turn produces heavier elements, including neon and silicon. The various nuclear reactions follow each other at an ever-increasing pace. Due to the enormous energy and overwhelming radiation pressure, the star is nearly torn to pieces. But then, when iron and nickel atoms have formed in the nucleus of the star, the spontaneous fusion reactions come to an end and the star explodes in a brilliant finale.

The supernova eruption throws most of the stellar gas into space. For days to weeks, the dying heavyweight shines as brightly as billions of suns. Its planets evaporate like snow in the sun; neighboring stars are blasted by a radioactive cloud that races through space at a speed of tens of thousands of miles per second.

The hot, expanding gas shell, pregnant with heavier metals and rare elements, marks, for thousands of years, the place where a short-lived giant star theatrically ended its life. Is there anything left of this

hapless star? Yes, there is: at the center of the remnants of the supernova is a small, super-compacted neutron star – the collapsed core of the star, heavier than the Sun but no bigger than a city like Boston or Atlanta.

Densely stacked nuclear particles – uncharged neutrons – give the mortal remains of a star an unimaginable density of 1.6 billion tons per cu. inch (100 million tons/cm^3). The neutron star rotates at the speed of a drill and the extreme magnetic field whips away bundles of high-energy radiation. The exotic celestial body thereby sometimes reveals its existence to radio telescopes on Earth: they detect a fast, flashing radiation source in the sky – a pulsar.

But the play isn't quite over yet. In the last act, a neutron star can absorb gas from a companion or collide with a companion after the second star of a binary system has also gone through a supernova explosion. A neutron star collision of this nature shakes space-time to its foundations. Even hundreds of millions of light-years away, sensitive detectors are able to capture minute gravitational waves. In a fraction of a second, the amount of energy the Sun produces in 100,000 years is released. In a new outburst of nuclear reactions, gigantic quantities of heavy metals are produced, including gold and platinum.

When the mortal remains of an exploded star (or the result of the collision and merger of two neutron stars) is twice as heavy as the Sun or more, even the densely packed neutrons are no longer immune to the relentless grip of gravity. In a catastrophic gravitational collapse, the matter collapses to a single point, and the rest of the star disappears forever from the stage, withdrawn from view by an inextricable knot of space and time – the horizon of a black hole.

But while supernova explosions destroy planets, neutron stars pump deadly X-rays into space and black holes devour everything under their gravitational influence, the birth mechanisms of the Milky Way never stand still. Ejected gas clouds contract again elsewhere to form new protostars, dust particles become planets and in the dark catacombs of molecular clouds organic molecules patiently wait for their chance. They then rain into a shallow, warm pool on the surface of a newborn planet, transforming chemistry into biology. Some day, the stardust will come to life.

The black glutton

A black hole (on the right in this image) absorbs gas from an accompanying star. The gas collects in a flattened rotating disk before disappearing behind the horizon of the black hole. The X-ray radiation of this hot "accretion disk" reveals that a black hole is present, even if it doesn't emit any electromagnetic radiation itself.

The Center of the Milky Way

Here's a delightful tale: two young astronomers from Leiden lost sight of their professor during their visit to South Africa in 1952 while carrying out test measurements for a new telescope in a dark place. This university teacher was Jan Oort, one of the greatest astronomers of the last century and a pioneer in the exploration of the Milky Way. The 52-year-old professor was behind a small hill lying on his back in the grass. In the pitch blackness and all to himself, he enjoyed the spectacular sight of the Milky Way stretching from horizon to horizon, with the mysterious galactic center high in the sky. Mysterious, as not much can be seen of the center of the Milky Way (situated in the border area of the constellations Scorpio and Sagittarius) from Earth.

Although the Milky Way band is wider and lighter here than elsewhere, it's also crossed and partly darkened by thick clouds of dust. The Indigenous peoples of South America and Australia recognized a jaguar, a llama and a giant emu in these bizarre cloud formations – dark constellations that can only be seen from dark places. (This part of the Milky Way is not visible from North America, but dark clouds can also be seen in the constellation Cygnus.)

No one would have thought that behind this thick curtain of dust was a dazzling splendor of stars. At the beginning of the 20th century, little was known about the absorptive effect of the Milky Way's dust. Oort's teacher Jacobus Kapteyn was therefore convinced that the Milky Way was relatively small and that the Sun and Earth were not too far from its center – comparable to the impression of a small, self-contained world that could be experienced on a foggy evening. Oort's research into the movement of the stars, however, clearly showed that the Milky Way, with the Sun about 30,000 light-years away from the center, is much larger than his former teacher Kapteyn had suspected.

It wasn't long before the actual size and majestic spiral structure of the Milky Way system were mapped, but not using optical telescopes but rather through the brand-new field of radio astronomy. By 1931, it had already been discovered that radio waves were coming from the center of the Milky Way. At the end of the 1950s, Oort and his colleagues had succeeded in reconstructing a "map" of the Milky Way based on measurements of the velocities of gas clouds. So we came to know our cosmic residence as a colossal spiral system with a diameter of about 100,000 light-years, the Sun being one of many billions of stars located somewhere in the quiet outskirts.

The larger and more sensitive radio telescopes became, the more the mysterious center of the galaxy revealed its secrets. In 1974, an extremely compact source of powerful radio waves was discovered in the constellation Sagittarius: Sagittarius A*. It measures less than 40 microarcseconds in diameter in the sky. In terms of the distance from the center of the Milky Way (27,000 light-years), this corresponds to an extent of over 30 million miles (50 million km) – only a third of the distance between Earth and the Sun. The ultra-compact radio source also seemed to coincide exactly with the center of gravity of the Milky Way. Could Sagittarius A* really be a supermassive black hole surrounded by a rotating disk of gas and dust that generates radio waves and X-rays?

Twenty years ago, this was still pure speculation. Today, however, there is no doubt about it. Using sensitive infrared telescopes capable of seeing through the Milky Way's dust, astronomers have witnessed heavy giant stars circling around the radio source in small orbits at unimaginable speeds.

One of these stars, S2, completes its long orbit in only 15.2 years, approaching Sagittarius A* at a

Gravitational spiral

Since the middle of the last century, we've known that the Milky Way is a gigantic spiral system with a diameter of around 100,000 light-years. The Sun is about halfway between the center and outer edge of the spiral, on the inside of a small spiral arm. This illustration is based to a considerable extent on infrared measurements taken by NASA's Spitzer Space Telescope.

A dusty spectacle

The bright center of the Milky Way, about 27,000 light-years from Earth, is obscured from view by the absorbing influence of bizarre dark dust clouds. This panoramic photo also shows numerous bright star-forming regions, including the Rho Ophiuchi cloud (on the right), which is "only" 450 light-years from Earth.

Unobstructed view

The central region of our galaxy, observed with infrared (red), near infrared (yellow) and X-ray wavelengths (blue). These types of radiation, which are invisible to us, aren't impeded by interstellar dust, or at least much less so. The radio source Sagittarius A* (the real center of the Milky Way) is located in the bright area at the right of the center of the photo.

Zoom into the center

This image of the bright giant stars near Sagittarius A* was taken with a sensitive infrared camera at the ESO's Very Large Telescope observatory in Chile. Sagitaruis A* is a black hole at the center of the Milky Way (it is not visible in this photo). Astronomers were able to deduce from the movements of the stars that the black hole is about four million times as heavy as the Sun.

distance of 11 billion miles (18 billion km) from us. Based on the trajectory of the swirling stars, it's easy to deduce that there must be a mysterious object in the center that weighs about 4 million times more than the Sun. It can't be an extremely compact star cluster, which would also be visible by infrared light. Moreover, star clusters are not known to emit radio waves and high-energy X-rays. This means that there's only one possible conclusion: the center of the Milky Way is home to a gigantic black hole with a diameter of about 15 million miles (25 million km).

Black holes are structures in space in which the gravitational force is so strong that nothing can escape them. The curvature of space-time – Albert Einstein's way of describing gravitational force – is so extreme that even a ray of light can't escape the attraction of a black hole. The cosmic glutton reveals its existence only through the influence of gravity that it exerts on its surroundings, as with whirling stars.

At the same time, a considerable amount of radiation is generated in the immediate vicinity of the black hole. Absorbed material first piles up in the circular accretion disk. The hot gas in this disk then emit high-energy radiation before disappearing behind the "horizon" of the black hole. The great distance and the dense, absorbent dust clouds unfortunately make it very difficult to observe the black heart of the Milky Way in detail. At this distance, only the brightest stars can be seen; the thousands and thousands of weaker specimens remain invisible.

X-ray measurements have shown that there are likely to be many thousands of smaller black holes in

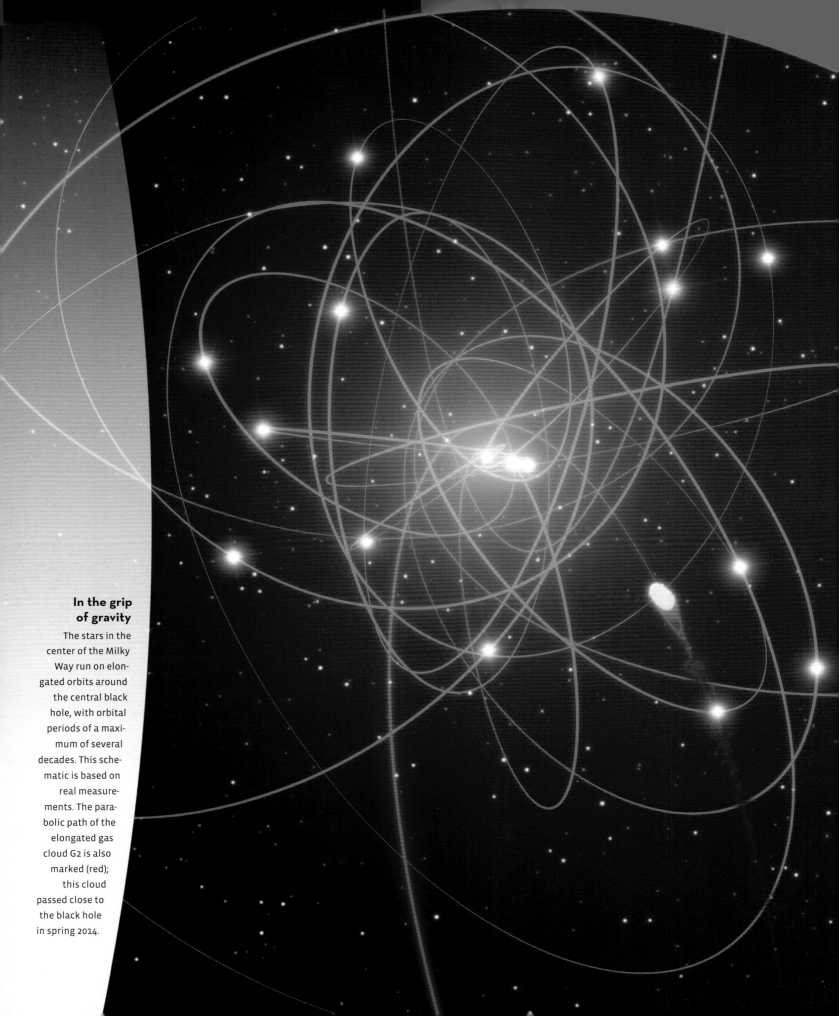

In the grip of gravity
The stars in the center of the Milky Way run on elongated orbits around the central black hole, with orbital periods of a maximum of several decades. This schematic is based on real measurements. The parabolic path of the elongated gas cloud G2 is also marked (red); this cloud passed close to the black hole in spring 2014.

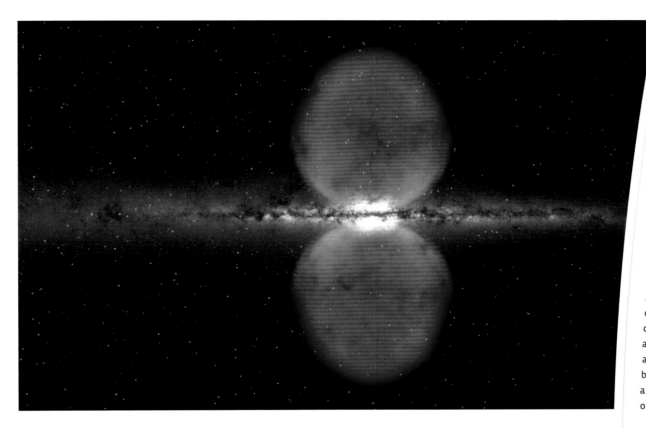

Galactic burp

In 2010, the American space telescope Fermi discovered huge "bubbles" of high-energy gamma radiation that extend up to 25,000 light-years above and below the center of the Milky Way. They must have been created by a massive explosion in the center of the galaxy about 6 million years ago, when the central black hole devoured a huge amount of matter.

the center of the Milky Way, but we don't know if neutron stars and pulsars are also hidden there. That is unfortunate, as precision measurements on cosmic metronomes of this kind provide valuable information on the gravitational field there.

Similarly, a long gas cloud that passed nearby in the spring of 2014 failed to deliver the expected galactic fireworks from which valuable astronomical data could have been obtained. What is certain, however, is that the large black hole at the center of the Milky Way regularly indulges in rather large cosmic meals. The amount of X-ray radiation emitted by Sagittarius A* varies greatly; the most violent eruptions are likely to occur when a huge amount of matter is absorbed within a short period of time – perhaps in the form of a cosmic boulder dozens of miles in size.

Sometime in the middle of the 17th century, the Milky Way's central black hole must have eaten a very large meal: the energy-rich radiation released during that process has now reached the molecular cloud at a distance of 350 light-years from the center of the galaxy. As a result, the gas in this cloud emits X-rays.

Even more spectacular was an eruption six million years ago, around the time when humans' distant ancestors practised walking upright for the first time. The center of the Milky Way released enormous amounts of gas and radiation in two directions perpendicular to the flat disk of the system.

In 2010, gamma-ray detectors orbiting around Earth were able to detect "bubbles" generated by this release of gas and radiation, which extend up to 25,000 light-years above and below the Milky Way plane. Astronomers also discovered compact clouds of hydrogen gas that were blown into space at high speed.

However, six million years is a blink of an eye considering the age of the Milky Way; no one knows when Sagittarius A* will be ready for its next large meal. Meanwhile, radio astronomers all over the world have joined forces to finally "capture" the supermassive black hole in the center of our galaxy on camera.

Parabolic antennas in Europe, the United States, Mexico, Hawaii, Chile and even the South Pole are (electronically) interconnected to provide the clearest image possible. With a little luck, this virtual telescope, referred to as the Event Horizon Telescope, can succeed in actually imaging the horizon of the black hole (the famous black hole photograph taken in April 2019 is of the black hole M87, near Messier 87).

Jan Oort could only dream of such insights, results and expectations at the beginning of the 1950s. It's thanks to large optical telescopes and radio antennas on Earth, as well as astronomical research into X-rays and gamma rays, that our knowledge of the Milky Way has grown unimaginably in just a few decades.

By no means have we solved all the puzzles – we're far from it. But even if we can only see it from the inside, of the hundreds of billions of galaxies in our perceivable universe, our own Milky Way is certainly most familiar to us, as it is our cosmic home.

INTERMISSION

Surveying the Milky Way

In the second century BCE, Greek astronomer Hipparchos compiled the first star catalog: a list of positions and magnitudes of about 850 stars in the sky. The European Space Agency's Gaia space observatory is adding a little more to it. Gaia is measuring the position in the sky of a billion stars of the Milky Way, as well as the distance, spatial motion, color and brightness of several hundred million stars. Gaia's measurements are extremely accurate, which means that even small fluctuations caused by orbiting planets are noticeable. Gaia also sees planetoids, ice dwarfs, comets, variable stars, supernovas, galaxies and quasars. Never before has the Milky Way been captured in such detail – in three dimensions. Gaia was launched at the end of 2013 and measurements are expected to be completed in 2019.

A cloud full of baby stars

Gas nebulae glow due to the ultraviolet radiation of newborn giant stars, some of which are covered in dust veils. This spectacular scene took place in the star-forming region N159 in the Large Magellanic Cloud, a relatively small galaxy about 160,000 light-years away from the Milky Way. N159 has a diameter of 150 light-years.

Cosmic Neighbors

The Magellanic Clouds

Persian astronomer Abd ar-Rahman as-Sufi never saw the blurry nebula himself. He was at the court of Emir Adud ad-Daula in Isfahan, Persia, about 215 miles (350 km) south of today's Iranian capital, Tehran. Isfahan lies at 32 degrees latitude north, so as-Sufi could not observe most of the southern sky. The Large Magellanic Cloud, for instance, never climbed above the horizon. He had, however, heard stories from mariners about a pale white spot between stars that could be seen from the Arabian Sea and stands deep in the southern sky. In his beautifully illustrated *Book of Fixed Stars*, which first appeared in 964 CE, as-Sufi described the nebula as *al-Baqar al-abyad* – the White Bull.

The blurry speck of light and its smaller companion have surely attracted attention since time immemorial. But when *Homo sapiens* began to spread across the globe from Africa more than 100,000 years ago, their collective stories were lost. And so the indigenous peoples of South America, Africa, Southeast Asia and Australia all created their own myths about the two nebulae that look like displaced highlights of the Milky Way. The Babylonians, Egyptians, Greeks and Persians, however, had no idea it even existed.

Ferdinand Magellan, the Portuguese explorer who made the very first trip around the world at the beginning of the 16th century, probably didn't know about as-Sufi's book. In 1519, with five ships and a crew of 270, he left Seville for the west. Tree years

Bridging

Despite their great distance of tens of thousands of light-years from each other, the two Magellanic Clouds are connected by a thin "bridge" of stars (difficult to see here) and neutral hydrogen gas. The distribution of this gas (reproduced in blue) was recorded using a radio telescope. This "Magellanic Current" was probably created by tidal forces.

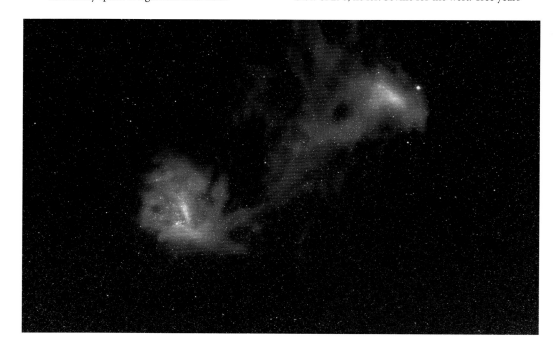

Southern attractions

The Large and Small Magellanic Clouds are not visible from most of the Earth's northern hemisphere. They are, however, as familiar to inhabitants of the southern hemisphere as the Big Dipper is to us in the northern hemisphere. Here you can see the two small companions of our Milky Way above the parabolic antennas of the ALMA observatory at 16,400 feet (5,000 m) altitude in northern Chile.

Milky Way companion

The Large Magellanic Cloud is about one-tenth the size of the Milky Way. Scientists hypothesize that it's a barred spiral galaxy that has been deformed by the tidal forces of the Milky Way. The brightest star-forming region, at the upper edge of the galaxy, is the Tarantula Nebula. This excellent overview photo was taken with a digital camera and a telephoto lens.

later, the *Victoria* was the only ship to return from the east with 18 survivors on board. One of those survivors was the Italian scholar Antonio Pigafetta, who recorded the first western observations of the two nebulae in the southern sky in his journals.

In 1603, German celestial cartographer Johann Bayer included the "Nubecula Major" and the "Nubecula Minor" (the large cloud and small cloud) in his monumental star atlas *Uranometria*. Later, Dutch sailors called them the Cape Clouds, after the Cape of Good Hope. Today, however, they are known as the Magellanic Clouds. For inhabitants of the tropics and southern hemisphere, they are just as well known and familiar as the Big Dipper and Polaris (the North Star) in the north.

During the 19th century, when European and American astronomers shipped some of their telescopes to more southern locations, it was discovered that the Magellanic Clouds are not nebulae but rather – just like the Milky Way band itself – made up of countless individual stars. They are obviously quite small compared to our own galaxy (which at that time was still considered the only one of its kind).

The fact that there are many billions of galaxies in the universe is a realization that only gained popularity at the beginning of the 20th century. Today we know that the Large Magellanic Cloud is a small, deformed barred spiral galaxy with an elongated center. The structure is about 14,000 light-years in size – one seventh the size of our galaxy – and is estimated to have tens of billions of stars. The Large Magellanic Cloud is 163,000 light-years away from us.

The Little Magellanic Cloud is slightly further away, at about 200,000 light-years, and is roughly half the size of its big sister. Sensitive telescopes have found a long "bridge" of weak stars between the two galaxies. In addition, they are both enveloped in a very large but very thin cloud of hydrogen gas. All this suggests that the two galaxies are exposed to strong tidal forces, both from the Milky Way and from each other, which is causing them to slowly but constantly drift apart.

Detailed investigations on the light of the Magellanic stars show that they contain less heavy elements than the stars of the Milky Way. In this respect, the two small, somewhat irregularly shaped clouds resemble the very first galaxies that saw light shortly after the Big Bang. They also consist almost entirely of the two lightest elements in nature, hydrogen and

Native soil

Through a telescope, we can see that the Large Magellanic Cloud contains countless cosmic nurseries – the birthplaces of new stars. The largest is the Tarantula Nebula (at the top of this photo), which contains several hundred thousand young stars. Under the Tarantula Nebula (at left), there are three regions where red stars are formed: N158, N160 and N159.

helium. In a large galaxy like ours, the original chemical composition was disturbed long ago by "contamination" from the products of nuclear fusion from earlier generations of stars, including carbon, oxygen and sulfur.

Stars also form in these two small galaxies. The Large Magellanic Cloud is home to the largest and most active star-forming region in the general area of the Milky Way: the Tarantula Nebula. The name refers to the structure of the nebula. Its gas veil stretches out like rays and is very reminiscent of the legs of a giant spider. With a diameter of some 600 light-years, the Tarantula Nebula is so large and bright that it is easily visible to the naked eye, despite its immense distance from Earth. If this star factory were located near the Orion Nebula in our Milky Way, it would be bigger than the constellation Orion, and it would considerably brighten the entire night sky.

The largest star cluster in the Tarantula Nebula, NGC 2070, is estimated to have nearly half a million young stars. Deep in the center of this cosmic collection are extremely massive stars that are less than two million years old. The object with the designation R136 ("R" stands for Radcliffe Observatory) was initially thought to be a giant star, but thanks to the sharp view of large telescopes on Earth and in space, we now know that several hundred stars are densely packed here in an area with a diameter of less than 30 light-years. The heaviest object, R136a1, weighs about 300 times more than the Sun and is the heaviest star astronomers have ever discovered.

But it won't be long before R136a1 ends its short life in a gigantic supernova explosion. The same fate awaits most other giant stars in the cluster. Elsewhere in the Large Magellanic Cloud, where some stars have already reached a greater age, these explosions have already taken place. The last one happened in February 1987, on the edge of the Tarantula Nebula. Supernova 1987A was easily seen from Earth with the naked eye and is the greatest stellar explosion ever observed. It is only a question of time until another star ceases to exist in the "White Bull."

Compared to its larger companion, the Small Magellanic Cloud is much less impressive. It plays second fiddle in almost every respect. It's slightly smaller (and farther away); it contains much fewer stars; its content of heavy elements is even lower than that of the Large Cloud; and the star-forming regions are more modest in size and less active. However, more neutron stars – the rapidly spinning remains of exploded stars – were discovered in the Small Magellanic Cloud. It also has more X-ray binaries, which are systems in which a neutron star (or stellar black hole) orbits around a heavy "normal" star. It seems

Heavyweights

The light blue stars in this Hubble photo belong to the open cluster NGC 2070. At the center is the heaviest known star in space: R136a1, which is about 300 times heavier than the Sun. The formation of such giant stars is likely made possible by the gas in the Large Magellanic Cloud, which contains relatively few heavy elements.

Embryonic stars

NGC 346 is one of the star-forming regions within the Small Magellanic Cloud. In addition to gas nebulae, dust veils and young giant stars, the Hubble Space Telescope has revealed newborn protostars in which nuclear fusion reactions of hydrogen are only just beginning. Some of these stars are only half as heavy as our Sun.

Small neighbor

The Small Magellanic Cloud, located 200,000 light-years from Earth, appears to be less active than its big sister. Nevertheless, there are numerous star-forming regions here as well. At the top of the photo you can see the huge globular cluster 47 Tucanae. It is not, however, part of the small, irregular galaxy but is instead 15,000 light-years away.

both clouds have undergone quite different evolutions.

The Small Magellanic Cloud made history at the beginning of the 20th century, when American astronomer Henrietta Swan Leavitt explored variable stars in the galaxy. She discovered that in a certain type of variable star – the so-called Cepheids – there is a close relationship between the rate at which the brightness of the star changes and the actual luminosity of the star. This is called the period-luminosity relationship, and it is an important pillar for determining distances in space: if you know how much light a star emits, you can deduce its distance based on the observed brightness in the sky.

Just like the center of our own Milky Way, the Large and Small Magellanic Clouds remain rewarding objects of study for astronomers as they are the closest neighbors of the Milky Way. The Magellanic Clouds are why new large telescopes are generally being constructed in the southern hemisphere. Time and again, this research delivers surprising results. Precision measurements taken by the Hubble Space Telescope have revealed that the two satellite galaxies are moving faster than previously thought; perhaps even so fast that one day they'll no longer be able to hold on to the gravity of the Milky Way. If "Nubecula Major" and "Nubecula Minor" are indeed unique visitors – accidental passersby on the cosmic stage and not permanent partners of the Milky Way – then we are lucky to be able to witness their visit and benefit from their proximity.

The Andromeda Galaxy

It's a cold, moonless autumn evening. Far from populated areas and the disturbing light pollution of our civilization, I gaze at the starry sky. Ursa Major – the well-known constellation whose brightest stars form the Big Dipper – stands low in the northern sky. To the west, the summer stars Deneb and Vega can still be seen; to the east, the winter constellations Taurus, Gemini and Orion rise. The Milky Way runs as a cloudy trail from east to west, crossing the zenith. In the middle of the Milky Way sits the W-shaped constellation Cassiopeia. High above the southern horizon, I look for the Andromeda Galaxy.

I know exactly where to look for it. From the upper left corner of the Great Square of Pegasus, I go two stars to the left and then two stars to the top. Yes, there it is: a blurred little spot of light in the sky, especially noticeable when I let my gaze wander a bit past it. My binoculars show a much clearer picture: an elongated nebula with a remarkably bright center. I, of course, know the Andromeda Galaxy from photos taken with a large telescope. Compared to that, this spot of light is nothing. What is impressive, however, is the fact that I am looking at starlight emitted 2.5 million years ago. I have my eye on another galaxy – the big sister of our own Milky Way.

The Andromeda Galaxy. It was undoubtedly known in ancient times, as the Persian astronomer as-Sufi described it. Even after the invention of the telescope, the Andromeda Nebula, as it was then known, retained its importance. What was this mysterious nebula that also seemed to have a beautiful symmetrical spiral structure? Was it a cosmic vortex – a swirling cloud of gas that would eventually form a new star? And would that happen to all those other spiral nebulae discovered during the 18th and 19th centuries? At what distance from Earth was this spot of light actually located? And how big could it be?

In 1885, a new star did indeed flare up in the Andromeda Nebula, although not in the center. Was it a so-called nova? Given the brightness of the new star, it was surmised that the nebula is part of our own Milky Way and possibly a complete star cluster in the making. But what about the much weaker "new stars" that were found later? Had these newly

A dusty mosaic
More than 3,000 single infrared photos of the Andromeda Galaxy were combined to create this mosaic. The photos were taken by NASA's Spitzer Space Telescope. The dust in the galaxy can be seen particularly well in infrared wavelengths (the "thermal radiation"), which is shown here in red. The blue colors show the distribution of predominantly old stars in the galaxy.

Big sister

The Andromeda Galaxy is the closest big sister to our Milky Way. From Earth, we see the spiral system at a flat angle. The Andromeda Galaxy is larger and heavier than the Milky Way and contains more stars. Despite being incredibly far away from us, at a distance of 2.5 million light-years, the galaxy can be seen with the naked eye on a clear autumn night as a small, elongated spot of light in the sky.

Diving into Andromeda

Hundreds of millions of stars and thousands of star clusters can be seen in this mosaic of the Andromeda Galaxy created from Hubble photos. With one and a half billion pixels, it's the largest image ever taken by the space telescope. On the lower left you can see the center of the galaxy and on the right one of its large spiral arms. The photo spans an area of about 60,000 light-years.

Energized stars

The Hubble Space Telescope captured thousands of bright blue stars deep in the center of M32, one of the elliptical companions of the Andromeda Galaxy. Based on the speeds at which these stars move around the center, astronomers have deduced that there must be a supermassive black hole in this region that is several million times heavier than our Sun.

discovered stars been nova explosions, the nebula would have been much further away and the explosion of 1885 would have been much more energetic. How could anyone ever figure this out?

How to precisely determine distance is a well-known problem in astronomy. For the stars that are relatively close to the Sun, we know the distances quite well. They show a small annual fluctuation in the sky – an apparent change in position caused by the Earth's movement around the Sun. The size of this "parallax" is a direct indication of the star's distance from any given object. Unfortunately, the effect of these fluctuations is always so small that it can only be accurately measured against the nearest stars. To measure distances of between objects that are far away from each other, we must use different methods. At the beginning of the 20th century, no one had any idea how to determine the distance to a nebula. In the case of the Andromeda Nebula, progress was made in the early 1920s. Using the giant 8¼-foot (2.5 m) Hooker Telescope on Mount Wilson, near Los Angeles, American astronomer Edwin Hubble succeeded in observing individual stars in the Andromeda Nebula. It wasn't long before everyone was convinced that this was a complete galaxy – an "island universe" (in Hubble's words) comparable to our own Milky Way. Today we prefer to call it the Andromeda *Galaxy* and not the Andromeda *Nebula*.

When Hubble discovered a Cepheid in the Andromeda Galaxy – a star with very characteristic and regular light variations – he was able to determine how far away the galaxy is from Earth. Henrietta Leavitt had determined a good 10 years earlier that there was a relationship between the speed of the light variations and the actual luminosity of these stars. Hubble measured the period of the Cepheid – the time the star takes to go through one light cycle. With the help of the Leavitt Law, he could then determine the actual luminosity of the star. By comparing these findings to the perceived brightness, it was easy to calculate the distance. Today we know that the Andromeda Galaxy is 2.5 million light-years from Earth. It's larger than our own galaxy and contains more stars: at least one trillion – i.e. 1,000 billion – stars shine from the Andromeda Galaxy. Like the Milky Way, the Andromeda Galaxy contains

Hot rings

Utraviolet images clearly show the ring-shaped structures in the Andromeda Galaxy, which can also be seen in infrared photos. The high-energy ultraviolet radiation mainly comes from young, hot stars. The photograph also shows the dust clouds in which new stars are born. This image is composed of 11 images taken by NASA's GALEX space telescope.

Peaceful companion

The second elliptical companion of the Andromeda Galaxy has the catalog designation M110 or NGC 205. The galaxy is about 15,000 light-years in size, similar to the Magellanic Clouds that accompany our Milky Way. Dark dust regions can be seen in this photo taken by an Austrian amateur astronomer. The galaxy also contains many young stars.

sparkling star clusters, luminous gas nebulae, dark dust clouds, active star-forming regions, old globular clusters, planetary nebulae, supernova remnants and rapidly rotating neutron stars. It is the closest large galaxy in the universe, the closest "adult" neighbor, of the Milky Way.

From Earth, we see the Andromeda Galaxy at a fairly flat angle; it looks like an elongated ellipse when viewed through binoculars.

Since we view the Andromeda Galaxy more or less at a flat angle, a lot of stellar light is absorbed by dust clouds located in the flat disk. This is quite unfortunate. If we were to view the Andromeda Galaxy directly from above, it would be a spectacular phenomenon in the firmament: a bright, nebulous spiral amidst the sparkling foreground stars of our Milky Way. We would also have discovered much sooner that a gigantic ring of brighter nebulae and young stars is located within the galaxy – in its beautiful spiral arms, to be exact.

The Andromeda Galaxy is not easy to spot using a simple "optical" telescope. However, observations in ultraviolet and infrared wavelengths allow us to see this ring clearly.

We now also know that there's a smaller ring of dust closer to the center and that the thin disk is slightly warped. All these structures are almost certainly the consequences of a collision a few billion years ago with another, smaller galaxy – probably one of the two elliptical companions of the Andromeda Galaxy. After all, this galaxy also shows similarities to our own "home": it's accompanied by two relatively large satellite galaxies.

And what about the center of the Andromeda Galaxy? Is there also a supermassive black hole similar to Sagittarius A*? Indeed there is, and this black hole is much heavier than ours: about 100 million times as heavy as the Sun. Strangely enough, a second bright "nucleus" was discovered only five light-years

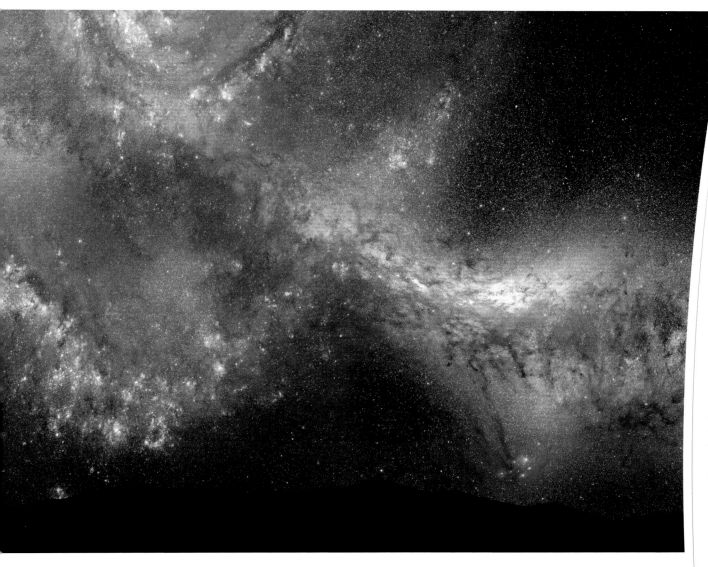

A head-on collision

The Andromeda Galaxy and Milky Way are approaching each other at a speed of about 70 miles per second (110 km/s). In a few billion years, the two galaxies will collide, be deformed by tidal forces and merge into a single large elliptical galaxy. From Earth, this will be a spectacular sight.

away from this black hole – possibly a large cluster of gas and stars kept on their path by the gravitational pull of the black hole. The Hubble Space Telescope (named after Edwin Hubble) only has a small field of view; this large nearby galaxy can't fit in a single Hubble photograph. But in recent years, dozens of photographs have been taken, which astronomers have used to create a gigantic mosaic. By comparing the razor-sharp photos with images taken several years ago, the space telescope was even able to measure the tiny positional changes of the stars in the sky. In other words, we can see that the Andromeda Galaxy gradually moves over time.

These precise measurements are important to determine the spatial motion of the galaxy. Studies of stars in the Andromeda Galaxy show that it is approaching our Milky Way at a speed of about 70 miles per second (110 km/s). This lateral movement was only recently discovered. We're now certain that the two large galaxies will collide in the distant future. In four to five billion years, the Andromeda Galaxy will merge with our Milky Way to form a huge elliptical galaxy, already known today as "Milkomeda."

So it seems that in a few billion years, I'll have to go outside again on a clear autumn night. The small, elongated spot of light will have long since grown into a frighteningly large spiral system that occupies half of the visible starry sky. Due to the mutual gravitational attraction between the Milky Way and the Andromeda Galaxy, the latter system will also be deformed and distorted, as will our own galaxy. Where gas clouds collide, thousands of bright new stars will emerge. At this point, Ursa Major and Cassiopeia will have long since disappeared. Incidentally, it's likely that mankind will also have disappeared. Our descendants will be history and all traces of our existence will have disintegrated into stardust. But that's a different story.

The Triangulum Galaxy

A constellation that's just a triangle? That sounds almost like a joke. With so many stars in the sky, you can form countless imaginary triangles. Ursa Major, Orion, Sagittarius, Crux (the Southern Cross) – these are all well-known constellations. But who has ever heard of a triangle?

Nevertheless, this small constellation is actually several thousand years old. It is, admittedly, rather inconspicuous, but it did attract some attention in ancient Greece. Greek astronomer Claudius Ptolemy, who lived and worked in Alexandria in the second century CE, included it in his famous list of 48 constellations. We still recognize all of these constellations today, even though the huge Argo Navis (the ship Argo) has been divided into three smaller constellations – Carina, Vela and Puppis (keel, sail and stern). The constellation Triangulum (Latin for "triangle") is a small elongated triangle of stars that sits

Home turf
One of the spiral arms of the Triangulum Galaxy contains the impressive star-forming region NGC 604: a gigantic gas and dust nebula in which new stars have been born for several million years. After the Tarantula Nebula in the Large Magellanic Cloud, NGC 604 is the largest cosmic delivery room in the "Local Group" of galaxies.

Radio radiation map

Radio telescopes in the Netherlands and United States were used to determine the distribution of cool hydrogen gas in the Triangulum Galaxy. The cool, dark gas isn't visible with a normal telescope. The radio radiation in this photo is represented in violet-blue colors. The spiral arms of the galaxy clearly extend further than one might expect at first sight.

Glowing arms
With a diameter of 60,000 light-years, the Triangulum Galaxy (M33) is much smaller than our Milky Way and the neighboring Andromeda Galaxy. Nevertheless, it steals the show in this overview photo, taken with the European Southern Observatory's VLT Survey Telescope in northern Chile. The pink spots in the galaxy's spiral arms are active star-forming regions; the brightest nebula is NGC 604.

between the constellations Andromeda and Aries. Like Andromeda, it's mainly visible during the fall. And just like the Andromeda Galaxy, a galaxy located in Triangulum is named after it. Thus, the constellation Triangulum is the eponym for the Triangulum Galaxy.

We don't, however, know if Ptolemy knew about this very faint nebula. Only if you have exceptionally good eyesight, are in pitch blackness and know exactly where to look can you hope to see the blurry spot with the naked eye.

This is simply sensational given its distance from Earth: three million light-years – about 18 trillion miles (30 trillion km). The galaxy can be seen quite well with binoculars, even if you don't see much more than a nearly round, blurry spot of light.

The Triangulum Galaxy is also known under the catalog designation M33. It's the 33rd object on a list of dozens of nebulae and star clusters drawn up in the second half of the 18th century by French astronomer Charles Messier – hence the M. Using a relatively small 4-inch (10 cm) telescope, Messier set off from the center of Paris to hunt for comets. They also look like faint blurry spots of light, but they move slowly between the stars. In order to better identify new comets, Messier decided to catalog the "stationary" nebulae. The final version of the Messier catalog was published in 1781 and contained 103 objects. Later on, seven more objects were added, including M110, one of the companions of the Andromeda Galaxy.

Many of the objects we've encountered in this book also have Messier numbers. The Andromeda Galaxy is M31, the Orion Nebula is M42 and the the Eagle Nebula is M16. Number 1 in the catalog is the Crab Nebula; M45 denotes the Pleiades (the famous open star cluster in the constellation Taurus), and we already know the globular cluster M92. All Messier objects can be seen with a small amateur telescope.

A hundred years after Messier, astronomers had much larger instruments, and there was a need for a more detailed catalog. In 1888, Danish-Irish astronomer John Dreyer compiled the *New General Catalogue*, comprising several thousand objects. We've already mentioned some of these NGC designations. The Messier objects are, of course, also included in the *New General Catalogue*: The Andromeda Galaxy (M31) is also known as NGC 224, and the Triangulum Galaxy (M33) is known as NGC 598.

Like the Andromeda Galaxy and our Milky Way, M33 is a spiral galaxy, just slightly smaller. Its diameter is approximately 60,000 light-years, and it is estimated to have about 100 billion stars – only about one-fifth of the number of stars in the Milky Way. Viewed from Earth, M33 is diagonally behind the Andromeda Galaxy.

The Triangulum and Andromeda Galaxies are about 500,000 light-years away from each other; M33 may be a distant companion of M31. Just like the Large and Small Magellanic Clouds, these two galaxies are "connected" by a thin bridge of gas and stars. There is evidence that the Triangulum Galaxy passed close by the larger Andromeda Galaxy a few billion years ago.

Today we can closely observe and document such phenomena, but 100 years ago it was quite different. In the early 1920s, Dutch-American astronomer Adriaan van Maanen (who by the way suffered from nyctophobia – the fear of darkness!) was still firmly convinced that "spiral nebulae" such as M31 and M33 were gas vortices in our own Milky Way. He was also confident that he could prove the accuracy of his thesis. In 1911, Van Maanen received his doctorate from Utrecht University after his studies on the so-called proper motion of stars – the very small positional changes of stars in the sky due to their motion in space. In 1912, while employed by the Mount Wilson Observatory in California, he decided to perform similar measurements on the weak stars in a handful of spiral nebulae, including M33.

Van Maanen compared photos that had been taken many years apart. He measured the positions of a row of bright stars as accurately as possible, and there did appear to be a very slow rotation of the spiral nebulae, with a rotation period of about 100,000 years. If the nebulae were far outside the Milky Way, as many astronomers claimed, the stars would have to move almost as fast as light around the center of the system. Only one conclusion was possible: the nebula had to be part of the Milky Way and were thus much closer to Earth – only then would the measured rotation period be realistic.

However, we know today that Van Maanen was wrong. Yes, spiral galaxies are rotating, but their rotation period is about a few hundred million years. It wasn't until 2005 that radio astronomers succeeded in measuring the proper motions of stars in M33. They move at less than 30 microarcseconds per year – a distance comparable to the thickness of a human hair viewed at a distance of about 300 miles (500 km). Strangely enough, it was never really clear why Van Maanen failed, since he was known as an extraordinarily careful observer. Maybe it was a case of wishful thinking.

Just as with the Milky Way and the Andromeda Galaxy, the spiral arms of the Triangulum Galaxy contain dust clouds and gas nebulae in which new stars are born.

By far the largest and brightest star-forming region is NGC 604 – so striking that it has been assigned its own NGC number. After the Tarantula Nebula in the Large Magellanic Cloud, NGC 604 is the largest and most active

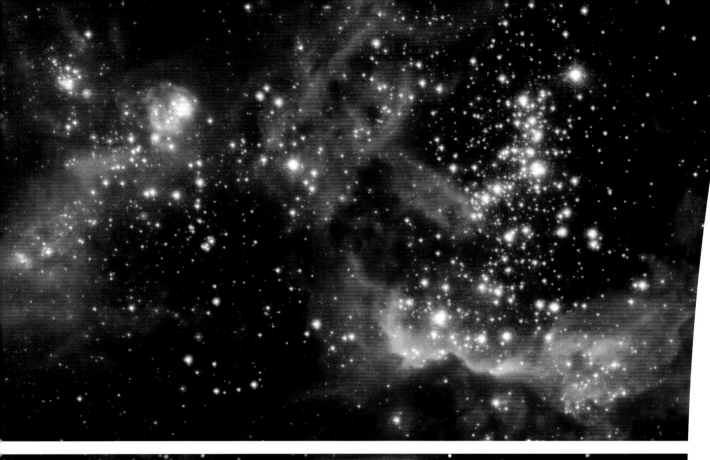

The cleaning process

In the center of the large star-forming region NGC 604, several hundred newborn hot stars emit energy-rich radiation that blows away the large cloud of gas and dust from within. In comparison, the Orion Nebula, a nearby star-forming region in the Milky Way, contains only four of these hot giant stars.

Thin and hot

Violent "stellar winds" and supernova explosions of stars within the center of NGC 604 blow large amounts of hot gas into space. The thin gas reaches temperatures of several million degrees and emits very high-energy X-rays. Observed by NASA's Chandra X-ray Observatory, this false-color image shows the X-ray radiation in blue.

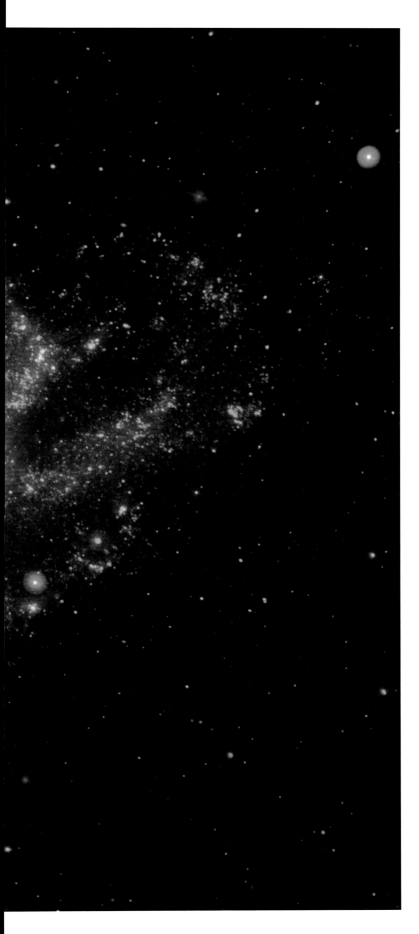

cosmic delivery room in nearby space, and it also includes a large young star cluster in the center.

Thus, despite its modest dimensions, M33 has much in common with its two large neighbors. But there's one important difference: measurements of the speed of movement of the stars in the center of the galaxy show that there's *no* supermassive black hole there. Although the center produces a large amount of high-energy X-ray radiation, if this radiation came from the direct vicinity of a black hole, it wouldn't be 10,000 times heavier than the Sun. This is quite remarkable, since in recent years it has become clear that almost every galaxy in space is home to a supermassive black hole.

However, lighter black holes, a consequence of supernova explosions, are abundant in the Triangulum Galaxy. They reveal their existence by suctioning gas from their environment. However, before the gas reaches the black hole, it's heated to such an extent that it emits X-rays. An X-ray source known as M33 X-7 ceases every three and a half days for a brief period of time: the black hole, which is about 15 times as heavy as the Sun, orbits around a gigantic star and is temporarily hidden from view once per orbit.

The Andromeda Galaxy, the Milky Way and the Triangulum Galaxy are the three largest galaxies in the so-called "Local Group." In addition to these three spiral galaxies, the Local Group also includes a large number of small dwarf galaxies, which are discussed in the following chapter. The Local Group, however, won't last forever. We've already established that the Andromeda Galaxy and Milky Way will collide in a few billion years. The fate of M33 is uncertain: it may participate in the galactic fusion, but it may also continue to circle around Milkomeda as a satellite galaxy.

Invisible light

Infrared and ultraviolet space telescopes expose details of the Triangulum Galaxy that are invisible or barely visible with optical telescopes. This photo shows the long-wave infrared radiation of dust clouds in the galaxy in red. The blue-green colors indicate the distribution of high-energy ultraviolet radiation emanating from young stars.

Satellite Galaxies

Three spiral galaxies spread over an area with a diameter of about 18 trillion miles (30 trillion km) – whose idea was it to call it the "Local Group"? Only astronomers could have gotten behind that, as astronomy isn't afraid of large numbers – 18 trillion miles is nothing compared to the dimensions of the observable universe. This galaxy group is indeed local: if Earth were to represent the universe, the Andromeda and Triangulum Galaxies would be in our village. Despite this neighborly relationship, new members that were previously unnoticed are constantly being discovered in the Local Group. Of course, these newly discovered galaxies aren't large spirals like our Milky Way, M31 and M33. They're small, inconspicuous dwarf galaxies, often home to no more than a few hundred thousand stars. They orbit the larger spiral galaxies like satellites at distances of several hundred thousand light-years.

They generally don't have bright young stars. A sensitive telescope is needed to detect such a thin cloud of weak stars from this distance.

Of course, the Large and Small Magellanic Clouds are also companions of the Milky Way. However, they are much larger and more striking, as are the two elliptical companions of the Andromeda Galaxy. In 1937, American astronomer Harlow Shapley found the first "dwarf satellite" of the Milky Way, a disorderly collection of stars in the constellation Sculptor. A year later, Shapley discovered a similar dwarf system in the constellation Fornax ("chemical furnace"), and in the 1950s four more small galaxies were added: two in Leo, one in Draco and one in Ursa Minor. When another one was found in 1977 in the constellation Carina, everyone spoke of the Milky Way and the Seven Dwarfs – with the Magellanic Clouds excluded so as not to mar such a catchy nickname.

Despite the tidy new nomenclature, researchers were confident that there were many more of these dwarf companions. In the 1960s, Italian astronomer Paolo Maffei discovered two quite large spiral galaxies in the middle of the Milky Way band at a distance of about ten million light-years from Earth (Maffei 1 and Maffei 2). The fact that they were previously undetected is attributable to the thick clouds of dust in the central area of the Milky Way. These galaxies are only clearly visible with an infrared telescope or radio telescope. The same applies to the two galaxies Dwingeloo 1 and Dwingeloo 2, which were discovered with the 82-foot (25 m) radio telescope of the same name in Drenthe.

The Maffei and Dwingeloo galaxies aren't part of the Local Group; instead, they belong to the next "village" in the cosmic landscape. However, if larger galaxies such as Maffei 1 and Maffei 2 can be obscured from our view by dust in the Milky Way, nearby dwarf galaxies can certainly be obscured.

One such inconspicuous companion of the Milky Way is the Sagittarius dwarf spheroidal galaxy in the constellation Sagittarius. It is about 50,000 light-years from us, and when viewed from Earth it is behind the center of the Milky Way. At the center of this small galaxy is the globular cluster M54, as well as the star clusters Palomar 12 and Terzan 7. Its center has a steep elliptical orbit, and its tidal forces are gradually pulling it apart. The path of the Sagittarius dwarf is scattered with countless weak stars originating from the galaxy. In a few billion years, the dwarf galaxy will have completely dissolved.

Many satellite galaxies whose stars are relatively far apart will suffer a similar fate: since the gravitational effect of the Milky Way is stronger on one side of the dwarf galaxy than on the other, it is constantly

---

A dwarf in the chemical furnace

The Fornax dwarf galaxy in the southern constellation Fornax is nothing more than a loose grouping of weak small stars. This small satellite galaxy – one of the many companions of our own Milky Way – was discovered in the 1930s by American astronomer Harlow Shapley. It consists of tens of millions of stars and at least six globular clusters.

Heart and soul

In certain infrared wavelengths, astronomers can look right through dust clouds of the Milky Way. This infrared image of the so-called Heart and Soul Nebula in the constellation Cassiopeia was taken by NASA's WISE satellite. The blue spiral galaxies in the bottom center of the photo are Maffei 1 and Maffei 2, which are hardly visible with a normal telescope.

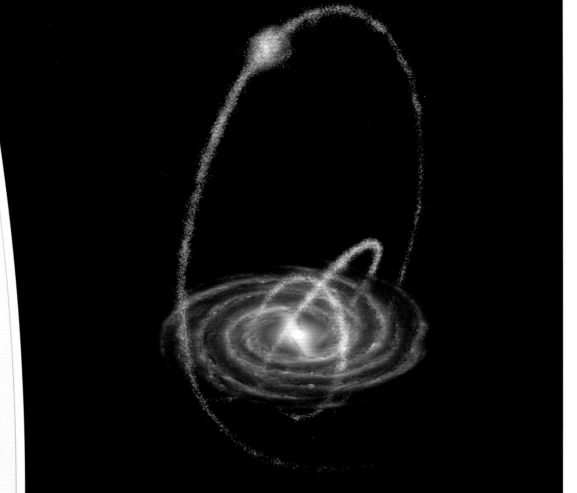

Pulled to pieces

Dwarf galaxies and star clusters that revolve in elliptical orbits around the Milky Way are slowly pulled apart by tidal forces over time. The stars are then scattered around the orbit, creating long "stellar streams," as can be seen in this image. The Sagittarius dwarf galaxy was also pulled lengthwise in this way.

being pulled apart, creating a long star trail. This makes it difficult to recognize it as a satellite galaxy.

Compact star groups are far less influenced by tidal action; they're held together by their own gravity. The globular cluster M54, which is part of the Sagittarius dwarf spheroidal galaxy, is one of the most compact globular clusters we know. Its compactness has, in fact, allowed it to resist the destructive tidal forces of the Milky Way. The less compact star clusters Palomar 12 and Terzan 7 have not been able to effectively resist the Milky Way's tidal effects, and their position in the sky has therefore been significantly altered.

The largest and brightest globular cluster in the Milky Way is Omega Centauri. However, it may in fact be the remaining nucleus of a sprawling dwarf galaxy that was "swallowed" by the Milky Way a hundred million years ago. Omega Centauri is located a short distance of 16,000 light-years from Earth and has a diameter of about 150 light-years. The globular cluster, which contains close to ten million stars, has a somewhat flattened shape, which is unusual for objects of this kind. What's more, some of the stars are younger than others. These two properties suggest that Omega Centauri is not a typical globular cluster from the Milky Way, but perhaps the central part of a former Milky Way satellite.

We've been discovering dwarf satellites in the Milky Way since the beginning of the 20th century. However, these discoveries have not been made by astronomers bent over a long-exposed photographic plate with a magnifying glass – as was common practice at the time of Harlow Shapley – but rather by large automated observation campaigns like the Sloan Digital Sky Survey. The entire starry sky has been captured several times with the help of very sensitive digital cameras. Special computer algorithms then search all measurement data for conspicuous patterns and structures.

In the meantime, about 60 satellite galaxies of the Milky Way have been identified, and this number is expected to increase in the coming years. In addition, astronomers are also finding an increasing number of "stellar streams" – the fossil relics of dwarf galaxies that have been pulled apart. The first to be found were the gigantic streaks of faint stars from the

A young globular cluster

The globular cluster Terzan 7, which is roughly behind the center of the Milky Way when viewed from Earth, once belonged to the Sagittarius dwarf galaxy. Research on the individual stars has shown that the globular cluster is "only" 8 billion years old – much younger than the other globular clusters in the Milky Way.

Mega Omega

Omega Centauri is the largest, densest and brightest globular cluster in the Milky Way. It is visible to the naked eye from the southern hemisphere as a small, blurry spot of light. The globular cluster contains millions of stars. Omega Centauri is possibly the remaining core of a dwarf galaxy that was, over time, pulled apart and devoured by the Milky Way.

A simulated universe

Using supercomputers, astronomers can simulate the evolution of the universe and the formation of large galaxies. These computer simulations predict that spiral galaxies like the Milky Way must be surrounded by a massive swarm of hundreds or thousands of small dwarf galaxies and accumulations of dark matter – more than can actually be observed.

Sagittarius dwarf spheroidal galaxy. Since then, almost 20 such stellar streams have been found, thanks to large automated search programs. Stars within stellar streams can be identified by the fact that they generally travel through the Milky Way in the same direction and at the same speed, as well as by their comparable chemical composition.

Of course, our Milky Way isn't the only galaxy that is accompanied by a large number of dwarf galaxies. About 30 of these small satellites have also been found in the direct vicinity of the Andromeda Galaxy.

Astronomers have even tracked down stellar streams in the outer regions of M31. Everywhere in space the same scene has been taking place for billions of years: small dwarf satellites are slowly being pulled apart by the tidal forces of a central galaxy and finally becoming part of the large system. And so, the development of spiral galaxies such as the Milky Way and Andromeda Galaxy are never complete.

With the help of powerful supercomputers, astronomers can simulate the formation of galaxies such as the Milky Way. They draw on the latest cosmological discoveries, such as the existence of dark matter – one of the mysterious components of the universe that will be discussed later in this book. The advanced computer simulations, in which the billion-year history of the universe takes place in minutes, actually predict that galaxies grow by devouring accumulations of dark matter and small dwarf galaxies.

However, the actual universe doesn't seem to stick entirely to the predictions of computer simulations. Or, to put it more accurately, the theoretical models that serve as a starting point for the simulations are probably still missing some important details. Scientists therefore predict that large galaxies like our own Milky Way are surrounded by many hundreds or perhaps even thousands of small dwarf galaxies, many more than have actually been observed. Perhaps many of these "mini galaxies" are composed almost exclusively of invisible dark matter and have formed virtually no stars; this might explain why we cannot see them.

But there is another problem. The simulations show that the dwarf galaxies are arbitrarily distributed in a large "halo" around the central galaxy and also move in many directions. But in reality, the dwarf satellites of the Milky Way and Andromeda Galaxy behave in a more orderly manner: they are more or less on one plane, and most of them move in the same direction. While no one has yet found a satisfactory explanation for this, technological advances and more accurate methods for measuring are sure to help scientists resolve these contradictions.

INTERMISSION

How Far Away Is this Star?

In the 1920s, American astronomer Edwin Hubble succeeded for the first time in determining the distance of a star in the Andromeda Galaxy from Earth. Hubble discovered that the star changed its brightness in a characteristic way. These stars are called Cepheids, and there is an observable relationship between the speed of their light variations and their luminosity. By measuring the cycle of the Cepheid, Hubble determined the luminosity of the star; by comparing his observations of their luminosity, he was able to determine their distance from Earth. The Cepheid is located in the lower left corner of this detailed photograph of part of the Andromeda Galaxy taken by the Hubble Space Telescope (named after Edwin Hubble). Today we know that the Andromeda Galaxy is 2.5 million light-years away from us.

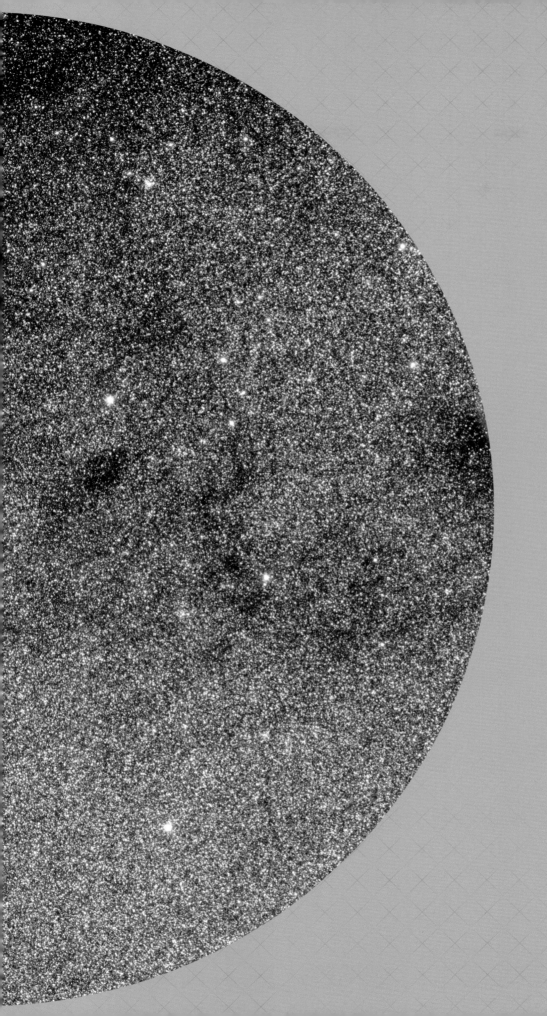

Majestic spirals
When viewed from Earth, the galaxy NGC 1232 is almost directly above us. The center consists mainly of old stars; the younger stars are in its majestic spiral arms. This galaxy is larger than our Milky Way and lies about 100 million light-years away in the constellation Eridanus. It has a slightly warped shape, which is the result of the gravitational pull of the small companion on its left.

A Gallery of Galaxies

Spiral Galaxies

The view from above

NGC 6814 is a wonderful example of a spiral galaxy. From Earth, we can see this galaxy almost directly "from above," giving us a good view of the bright center and the individual spiral arms with their dark dust nebulae and sparkling star clusters. This galaxy is located 75 million light-years from Earth, in the constellation Aquila.

My father used to have a book about the universe in his house. The title was *Das Welttall* ("the universe" in Dutch). It was from the famous Time Life book series. I could flip through it for hours and get lost in the breathtaking photos and illustrations. Somewhere in the back of the book, the size of the universe was explained with the help of a series of cubes. The first showed Earth circled by the orbit of the Moon. This cube was then scaled down 1,000 times and depicted in the center of a subsequent cube in which a large part of the solar system was drawn. In the next step, the Sun was shown as a small shining point in an empty room. Then the first stars appeared in the picture. Our Milky Way could be seen in the fifth cube. The sixth and final cube was reserved for the countless other galaxies in the space around us.

Even more impressive was the huge color photo of the Andromeda Galaxy, taken with the 16-foot (5 m) Hale Telescope at the Palomar Observatory in California, which also decorated the cover of the book. I envisioned being able to zoom in on the picture by repeatedly enlarging a small imaginary cube until I could finally land at one of the many billions of stars and see a small planet orbiting this inconspicuous star – a small twin brother of Earth, 2.5 million light-years away in the immeasurable universe. I still cannot suppress this tendency to search for Earth, so to speak, in other galaxies.

Distant galaxies, many tens of millions of light-years away from us, can now be captured in much greater detail by the Hubble Space Telescope than the nearby Andromeda Galaxy was captured by the earth-bound telescopes of half a century ago. The fascinating Hubble photos show gas nebulae, star clusters, dark dust clouds and, in many cases, even individual stars – you can spend hours looking at them. Time and time again, my gaze wanders to a point somewhere in the outer regions of such a galaxy, which corresponds to the position of the Sun and Earth in our own Milky Way. Look, there, invisibly small, that's where we are!

Irish nobleman William Parsons, better known as the Earl of Rosse, had the largest telescope in the world in the middle of the 19th century. It had a mirror made of shiny polished metal with a diameter of 6 feet (1.8 m). On exceptionally clear nights, the earl used it to explore the starry sky. He first discovered the spiral structure of some nebulae without realizing that he was looking at extremely distant "companions" of the Milky Way. In the case of M51, a blurry object located in the constellation Canes Venatici (under the tail of Ursa Major), the spiral structure was so striking that Lord Rosse called it the "Whirlpool Nebula." His masterful drawing of the whirlpool was depicted in that Time Life book.

Many other small flecks of light in the sky also seemed to be spiral nebulae. It wasn't until the 1920s, thanks to the work of Edwin Hubble, that it became clear that the Andromeda Nebula, the Whirlpool Nebula and their many spiral companions were separate galaxies located at great distances *outside* the Milky Way (and these systems are now known as the Andromeda Galaxy and the Whirlpool Galaxy). While many astronomers assumed that the Milky Way also had a spiral structure, it wasn't until the middle of the last century that Dutch radio astronomers recorded it for the first time.

It's logical: the street map of a city is easier to see when you fly over it in an airplane than when you're standing at street level. Likewise, the structure of

Open arms

Not all spiral galaxies look the same. NGC 5584, 70 million light-years from Earth, in the constellation Virgo, has loosely wound spiral arms. The galaxy was discovered in 1881 by American astronomer Edward Barnard. It played an important role in determining the distance scale of space, in part because a supernova explosion had recently occurred there.

Distorted spirals

The influence of gravity from two neighboring systems has slightly distorted the spiral galaxy M66. The spiral arms are asymmetric, and the bright nucleus is not exactly at the center. M66, which is 35 million light-years from Earth, in the constellation Leo, belongs to a group of three galaxies. It's clearly visible through an amateur telescope.

another spiral system, which we always look at from the outside, is much more visible than that of our own galaxy, which we can only view from inside.

There are plenty of spiral galaxies. Some have two noticeable arms, while others have four or more. In some instances, these spiral arms are very tightly wound around the center; in others, the structure is much looser and more open. And while many spiral galaxies have a high degree of order and symmetry, there are also galaxies with chaotic, distorted arms and a messy appearance.

Now for the dimensions: the Milky Way, with a diameter of 120,000 light-years, is one of the larger specimens. There are, however, spiral galaxies with diameters of a few hundred thousand light-years, while the smallest spirals are often no larger than 15,000 light-years. Most spiral galaxies have the same three-part structure. Their most striking feature is, of course, their flat shape and characteristic spiral arms. This disk contains mostly gas and dust; this is where you'll also find the dark molecular clouds and the bright gas nebulae in which new stars are born. Open star clusters and heavy, bright giant stars often seem to line up like glittering pearls along the spiral arms, while older stars spread over a slightly thicker disk over time.

In addition, the core of most spiral galaxies contains a compact collection of old stars in the form of a flattened sphere, the so-called central bulge. The stars here aren't very far apart and there isn't much interstellar gas. In photographs of distant spiral galaxies, the central bulge often looks like an overexposed "clump" of stars that, in line with their age, has a striking yellowish color. (White and blue stars are hotter and heavier than yellow and orange stars and have a much shorter life.)

Finally, spiral galaxies are surrounded by a more or less spherical halo in which there are also old, weak stars, but these are much further away from each other. The movements of the stars in the halo around the center of the galaxy are much more disordered than those of the stars in the disk. They move in all possible directions, usually along very long paths. The same applies to the galaxy's globular clusters, which are also located in the halo and are particularly concentrated toward the center.

Although a human life isn't long enough to notice any of this, each spiral galaxy has a slow rotation. The stars, dust clouds, gas nebulae and star clusters in the disk of the galaxy also all rotate in the same direction around the center. This usually happens at unimaginable speeds: our own Sun has an

The celestial windmill

This spiral galaxy (M101) is called the Pinwheel Galaxy because of its shape. It's relatively close to us – only 23 million light-years away, in the constellation Ursa Major. This highly detailed image is composed of no less than 51 individual images taken by the Hubble Space Telescope. The galaxy is about one and a half times the size of our Milky Way.

Sprinkled with stars

While some spiral galaxies appear to be wild and dynamic, NGC 2841 (located 65 million light-years from Earth, in the constellation Ursa Major) displays a pattern of cheerfulness and tranquility. The galaxy has remarkably short spiral arms that contain many dozens of star-forming regions, which resemble small, bright snowflakes when viewed at night. It's uncertain how such "flaky" galaxies are formed.

orbiting speed of 125 miles per second (200 km/s). However, since the Sun is 27,000 light-years away from the center of the Milky Way, it must travel 170,000 light-years in order to complete one orbit. Even at such a high speed, the journey will take about 250 million years.

Nevertheless, this doesn't mean that the rotation period of the Milky Way is 250 million years. Stars at a shorter distance from the center complete their orbit in a much shorter period of time, while stars at the outer edge of the galaxy take longer to complete their orbit. Other galaxies also show this phenomenon, called differential rotation: they don't rotate like a wagon wheel, but rather like our solar system, where the inner planets have a shorter rotation time than the outer ones.

So it is clear that spiral arms can't be rigid structures. If they were, they would have wound up tightly long ago based on the differential rotation of the galaxy. Instead, the spiral arms are more like a visible sequence of density waves propagating at their own slow pace through the galactic disk. The interstellar gas is compressed somewhat more strongly in the spiral arms, which makes it easier for new stars to form there.

Between each spiral arm, the density is lower and the stars are further away from each other. As a result, the stars in the disk of the galaxy move through the spiral arms, temporarily traveling through a region of higher density. This is roughly comparable to driving a car on a four-lane highway that suddenly becomes two lanes over a short distance. The cars then drive closer together on this narrower segment than before or afterward.

Much more can be said about the rotation of spiral galaxies, but we will save this for a later chapter in which the mystery of dark matter is discussed in detail. First we'll focus on different galaxies and astronomers' attempts to explain the many different varieties.

The overall impression

To understand a spiral galaxy properly, we need to look at it in infrared and ultraviolet wavelengths. In this false color image of NGC 3344, the infrared and ultraviolet lights better accentuate the star-forming regions and young star clusters. The bright star at the top left is in our own Milky Way; NGC 3344 is located 20 million light-years away from us.

Barred Spiral Galaxies

A test image

The barred spiral galaxy NGC 6217 was captured in 2009 by the Hubble Space Telescope's Advanced Camera for Surveys. The photos were taken to check whether the camera had been successfully repaired by NASA astronauts a few weeks earlier. In addition to visible light, the ACS also records infrared radiation, which provides detailed images of the star-forming regions within the galaxy's spiral arms.

Generally speaking, people like order. We want to see a system everywhere. This need to classify and categorize is also the basis for any form of science. In order to understand the underlying mechanisms and processes of nature, you have to begin by recognizing a pattern, find interrelations and then decipher correlations and causalities.

For this reason, astronomers will immediately draw a graph once they've measured two properties of a handful of similar objects. And for that reason, we try to reduce diversity in the universe to manageable pieces by dividing objects into easily understandable classes or groups.

But the cosmos doesn't always cooperate. Some planetoids have a tail of gas particles, which makes them look very similar to a comet. The question of whether Pluto is a dwarf planet or a "full-fledged" planet is still being debated. Brown dwarfs don't fit into the category of stars, but they don't fit into the grid of planets either. There is no clear distinction between globular clusters and dwarf galaxies – and so on.

Despite everything, if a new cosmic phenomenon is discovered, scientists are well advised to extract or give it a system and order it. What new insights such a classification ultimately brings is something we only care about later.

Edwin Hubble was the first to prove, in the 1920s, that spiral nebulae are actually independent galaxies, far outside our Milky Way. He also had the largest telescope of his time, the 8¼-foot (2.5 m) Hooker telescope at the Mount Wilson Observatory. With it, he photographed and studied as many of these systems as possible and, based on these observations, presented a classification system to the public in 1927: his famous "tuning fork diagram," also known as the Hubble sequence.

Initially, Hubble observed that spiral galaxies (S) are not always tightly wound. When the spiral arms of a galaxy were tightly wound around the core, Hubble gave it the type designation Sa. He gave slightly less narrow galaxies the designation Sb and the spirals with the most open structure the designation Sc. Simple but effective.

Hubble and his contemporaries also discovered that many spiral galaxies had an elongated center – a kind of bar-shaped band of stars. In a barred spiral galaxy (SB), the spiral arms don't start from the center, but rather from the ends of the bar. The galaxy, therefore, somewhat resembles a lawn sprinkler. These barred spiral galaxies are also not all tightly wound; Hubble therefore introduced the SBa, SBb and SBc classifications.

Furthermore, he observed many galaxies that didn't have any spiral arms at all. They resembled unstructured clusters of stars with the highest density in the center – a kind of oversized globular cluster. These were not exactly spherical but rather slightly elliptical. Based on this flattening, Hubble classified these galaxies from E0 (almost spherical) to E7 (very elliptical).

He himself was very pleased with his classification system: "Out of the thousand plus galaxies I have studied, only a dozen do not fit into any of these categories," he wrote.

Hubble placed the various classifications strategically in a diagram. He placed the elliptical galaxies from E0 to E7 on the left, on a horizontal line. He also referred to these as "early-type galaxies." To the right of E7, the line was split into two branches. The

A dusty bar

Dozens of Hubble photos were assembled into a spectacular mosaic of the barred spiral galaxy NGC 1300. The galaxy is located 60 million light-years from Earth, in the constellation Eridanus. The two long dust veils that extend from the center of the galaxy to the "base" of the two giant spiral arms are remarkable.

spiral galaxies (from Sa to Sc) were listed on the top branch, and the barred spiral galaxies (SBa to SBc) were on the bottom. The system and barred spiral galaxies were collectively referred to as the "late-type galaxies."

This system resembles a horizontal tuning fork, hence this graphic is often referred to as the tuning fork diagram. The terms "early-type" and "late-type" are, of course, quite suggestive. Although it was never Hubble's explicit intention, these designations suggest that the Hubble sequence is also evolutionary. A galaxy would begin as a spherical cluster of stars (E0), be flattened by rotation (to E7) and then either become a normal spiral galaxy (S) or a barred spiral galaxy (SB), whose arms would become less and less taut over time (a to c). This sounds quite logical, especially since elliptical galaxies mainly contain old stars, whereas spiral galaxies contain many young

Central heating

The central bar of NGC 4394, 55 million light-years from Earth, is less pronounced than that in other spiral bar galaxies. However, this photo clearly shows that the spiral arms don't originate from the center of the galaxy. Incidentally, there are large quantities of hot gas in the nucleus, although nobody knows for sure where the energy that causes such high temperatures comes from.

Intoxicating beauty

The impressive barred spiral galaxy NGC 1365, also known as the Great Barred Spiral Galaxy, sits 60 million light-years from Earth, in the constellation Fornax (chemical furnace), which can be viewed from the southern hemisphere. It's about twice the size of our Milky Way. This photo was taken with the Danish telescope at the European Southern Observatory's La Silla Observatory in Northern Chile.

stars. Soon after Hubble's classification system was published, lenticular galaxies were also discovered – a kind of transitional form between the elliptical and spiral galaxies. These were given the designation S0. In the Hubble sequence, they fit perfectly into the point where the tuning fork splits.

It is very tempting to interpret an observed pattern (in this case the different types of galaxies) as the result of evolution. Even before Hubble, astronomers classified the different stars in the sky on the basis of their so-called spectrum. They determined the spectrum by dividing the light of a star into the different colors of the rainbow, allowing them to see exactly how the energy production of the star is distributed over different wavelengths. Blue and white stars, the hottest stars, are almost always very bright; slightly cooler yellow stars like the Sun produce slightly less energy; orange and red stars, the coolest stars, are generally the weakest. Here, too, an evolutionary effect was initially assumed: a star would begin its life as a hot, bright, blue-white giant and over time cool and shrink to a red dwarf. Astronomers are still discussing the "early spectral classes" O, B and A (blue and white stars) and the "late spectral classes" F, G, K and M (orange and red stars).

We now know this isn't exactly how it works. Whether a star is born as a heavy, hot giant or as a light, cool dwarf depends primarily on the amount of gas in the collapsing cloud from which the star is formed. A star like the Sun doesn't just change color or spectral class: once a G star, always a G star. Although this isn't always the case either. At the end of its life, the Sun will swell to a red giant star and eventually shrink to a white dwarf. Something similar is true for galaxies. There is not, in fact, an evolutionary lifecycle as suggested by Hubble's tuning fork diagram A system that emerged as an Sb spiral may change type over time.

Elliptical, lenticular and irregular galaxies will be discussed in more detail in the next chapter; here we'll mainly look at how a normal spiral galaxy can become a barred spiral galaxy or vice versa. Astronomers haven't yet been able to answer this question, but it seems to have something to do with the elliptical orbits of the stars within the central part of the galaxy – the central bulge. A star in an elliptical orbit spends more time at a greater distance from the center than it does close to it. Fluctuations in density intensify this effect, which can result in an elongated bar that appears this for a long period of time.

We don't know for sure if a barred spiral galaxy also easily returns to a normal spiral. While there is some evidence that the bar at the center of spiral galaxies is a temporary structure with a lifetime of several billion years, it has also been found that the percentage of barred spiral galaxies is currently much higher than many billion years ago. Today, almost two out of three spiral galaxies have a bar at their center. However, when the universe was young, only one in five spiral galaxies had a bar.

The Milky Way also appears to be a barred spiral galaxy (of the SBb type). The central bar, however, can't be seen with simple telescopes, as the stars in the center are hidden from view by absorbing dust clouds. This isn't a problem for the NASA's Spitzer Space Telescope, which makes observations using infrared wavelengths. Research using the Spitzer telescope on the distribution of about 30 million stars in the central region of the Milky Way has revealed that there is a central bar that measures about 25,000 light-years in length, which we can view at an angle of about 45 degrees from our location in the outer regions of the galaxy.

The cosmic sprinkler

The spiral arms of NGC 1073 don't start at the nucleus of the galaxy but rather at the outermost ends of an elongated central bar of gas and stars. The galaxy therefore somewhat resembles an old-fashioned lawn sprinkler. Astronomers still do not know exactly how such barred spiral galaxies are formed. It's well known that they were much less common a long time ago and are more numerous today.

A bar in the eye
The tightly wound spiral arms of NGC 1398 are crossed by streaks of gas and dust. The central eye of the galaxy shows a short, bright bar; together, the innermost spiral arms form a noticeable ring. This fascinating galaxy is 65 million light-years from Earth, in the constellation Fornax (chemical furnace). This photo was taken by the European Southern Observatory's Very Large Telescope in Chile.

Ellipses, Lenses and Dwarfs

Astronomy is quite an extraordinary science. A geologist can take a piece of rock to a laboratory, where he or she can study it in myriad ways. A chemist can repeat a certain chemical reaction as often as he or she wants. A physicist throws elementary particles against each other and then looks at the result of this impact from every conceivable perspective.

An astronomer, however, must be content with what the cosmos has to offer. Nobody knows when the next supernova explosion will take place, and we can't ask the universe to repeat its last radio burst; round-the-clock observation is thus extremely important. Cosmic phenomena do their own thing, at their own pace. What's more, we can only see them from *one* angle.

Profile view of a lens

Since we are able to view the side of this galaxy almost perfectly from Earth, the dust clouds in the central plane of NGC 5866 can be seen very well: they appear dark against a giant elliptical halo of stars. Objects like NGC 5866 – a type of transition form between elliptical and spiral galaxies – are called lenticular galaxies.

A mixed duet
Astronomers don't know for sure whether these two galaxies are at exactly the same distance from Earth – there doesn't seem to be much interaction. The spiral system (NGC 4647) may be slightly further away or closer than the gigantic elliptical system M60. Both galaxies are part of the Virgo Cluster, which is about 50 million light-years from Earth, in the constellation Virgo.

That's caused a lot of confusion in the past. For example, you can calculate the total amount of energy created by a distant explosion in space by measuring how much radiation we receive on Earth and knowing the distance. In this kind of calculation, however, it's assumed that the explosion expands in every direction with the same force and brightness. If a distant celestial body by chance radiates light only in our direction, the total amount of energy is enormously overestimated. For some distant galaxies – called quasars, which will be discussed later in this book – this seems to be the case.

Even the spatial geometry of an astronomical object is sometimes difficult to determine with certainty. Stars are spherical, so they look more or less the same from every angle. The shape of a spiral galaxy in the sky, however, is determined by the angle at which we view it: perpendicular to the disk, as in a whirlpool galaxy, or diagonally from the side, as in the Andromeda Galaxy. In the first case, the galaxy appears almost round; in the second case we see a long nebula. The true three-dimensional shape of a elliptical galaxy is not always obvious.

The two conspicuous companions of the Andromeda Galaxy are beautiful examples of relatively small elliptical galaxies. They display no obvious central disk like our Milky Way and no spiral arms. At the beginning of the last century, Edwin Hubble and his contemporaries discovered many hundreds of such elliptical galaxies – unstructured aggregations of stars with a distinct concentration toward the center. Depending on their shape, Hubble gave them a type designation, from E0 (almost circular) to E7 (very elongated).

But how do we know what the actual shape of an E0 galaxy is? Maybe it's a large spherical cluster of stars that looks the same from every angle, but it could also be the top view of a rather flattened galaxy, which could have a shape similar to a bread roll. Or it could be the side view of a somewhat elongated galaxy shaped like a kiwifruit. And when we see an E3 galaxy (the most common type), is the perceived shape determined solely by the actual three-dimensional measurements, or does the angle at which we view the galaxy also matter? Using sensitive spectroscopes, astronomers can measure the velocity of stars in another galaxy. However, these types of measurements are generally not very useful in elliptical systems. Stars in this kind of system don't show clean, ordered movements like they do within the flat disk of a spiral galaxy. Instead, they move more or less in a crisscross pattern, as in the central bulges of most spiral galaxies. Measuring the speed inside elliptical galaxies is also not very helpful to decipher their three-dimensional structure.

However, measuring the volicity of stars provides valuable information about a completely different aspect of elliptical galaxies. The average orbital speed of the stars in all elliptical galaxies increases very rapidly as they move away from the center. In other words, the velocity distribution shows an enormous peak in the center of the galaxy. This can only be explained by a large amount of mass at this center that is compressed to a relatively small volume. Everything indicates that there are large, heavy black holes with a strong gravitational field there. When a supermassive black hole swallows large amounts of gas within the core of a galaxy, the gas is heated and emits high-energy X-rays before it disappears into the black hole. Many elliptical galaxies, however, are not noticeable sources of X-ray radiation, since they generally contain almost no interstellar gas. Molecular clouds and star-forming regions are also scarce there. There are no young, open star clusters or bright newborn stars. Stars in an elliptical galaxy are generally very old, giving the galaxy a yellowish color. It's quite unlike a spiral galaxy, which has a much bluer tint, created by young, hot stars, which have a life span of at most a few tens of millions of years.

Because of this aged appearance, scientists assumed that elliptical galaxies were among the oldest systems in space. Now, however, we know that when the universe was young there were fewer elliptical systems than there are today. It's obvious that more and more elliptical systems have been created during the course of cosmic history. Astronomers now believe they've discovered the process by which they are created: an elliptic system is the result of the collision of two smaller spiral galaxies. As mentioned previously, our own Milky Way will collide with the Andromeda Galaxy several billion years from now. Computer simulations suggest that this fusion will also lead to the formation of a new colossal elliptical system.

Elliptical galaxies thus don't owe their older radiation to their own age, but to the fact that they predominantly contain old stars that at some point belonged to two or more separate galaxies. As a direct result of their collision, these spiral galaxies appear to have lost most of their interstellar gas supply, resulting in essentially no new star formations in the remaining elliptical system. However, the details of this process are far from fully understood. It is also unknown whether all elliptical systems are the result

A magnetic effect

The elliptical galaxy NGC 1275, situated at the center of the Perseus Cluster, is partially obscured by a nearby spiral galaxy whose dark dust paths are particularly striking. The red chain-like structures with dimensions of many tens of thousands of light-years consist of relatively cool gas. They are held in place by strong magnetic fields in the galaxy.

The mysterious type

The Sombrero Galaxy (M104) in the constellation Virgo is 30 million light-years from Earth. We see the galaxy more or less from the side, making the dust clouds in the central plane clearly visible. We can't, however, observe whether or not the galaxy has spiral arms. The extensive halo of stars suggests that it's a lenticular galaxy.

An irregular dwarf

Huge bubbles of glowing hot hydrogen gas define this image of Holmberg II, a dwarf irregular galaxy located 11 million light-years from Earth, in the constellation Ursa Major. At the bottom left of the photo, the central part of the dwarf galaxy is visible. The largest star-forming regions are mainly in the outer edges, but new stars are also form elsewhere in the galaxy.

Deep effect
What separates the gigantic elliptical galaxy NGC 4696 from many other galaxies are its striking dusty tendrils, which are clearly visible in this Hubble image. The galaxy is 120 million light-years from Earth. In the background you can see countless galaxies that are much further away. The two bright stars are part of our Milky Way.

of such collisions. As for the gigantic specimens located at the center of large galaxy clusters – systems many hundreds of thousands of light-years in size that contain several trillion stars – there is little doubt about how they were created. But there are also elliptical dwarf galaxies measuring only a few thousand light-years in size, possibly from the earliest times of the universe.

Astronomers still don't know what to think about the strange lenticular galaxies that Hubble classified as S0. Just like spiral galaxies, they have a clearly recognizable central disk, which has an ordered rotation and in which – just like our galaxy – there are gas and dust clouds. However, there are no striking spiral arms and the central bulge is much larger than in regular spiral galaxies. Instead, it has the characteristics of an elliptical galaxy. Lenticular galaxies thus seem to represent a strange transitional form between elliptical and spiral galaxies, but it's not clear whether there's an evolutionary connection.

We still don't know the true nature of some galaxies. When we look at a galaxy from Earth more or less from the side, we can't really distinguish possible spiral arms anyway. We see the galaxy lying on its side, look at its central dust band and then often don't know whether it's a spiral galaxy with a relatively large central bulge or a lenticular galaxy. The well-known Sombrero Galaxy (M104), located in the constellation Virgo, for example, is described by some astronomers as a large spiral system with an extended halo of stars and by others as an S0 galaxy. Infrared measurements in turn suggest that the Sombrero Galaxy is actually an elliptical galaxy but with a prominent, dusty central disk.

Finally, there are also galaxies with bizarrely shaped outer sides without recognizable symmetry – called irregular galaxies – that don't seem to fit into any category. They sometimes display a lot of star-formation activity, and they probably have much in common with the very first small galaxies that formed about 13 billion years ago, shortly after the Big Bang. Hubble labeled these galaxies with the letters "Irr" – a catchall group for all galaxies that don't fit in with our penchant for order.

Dark Matter

"There are more things in heaven and earth, Horatio, than are dreamt of in your philosophy." This famous quote from *Hamlet* is one that I very much agree with. It would be boundless arrogance to claim that we know everything that exists. No matter how much science – our "wisdom" – has revealed to us, there are always things that are unknown. Shakespeare correctly recognized that.

This notion is entirely applicable to our knowledge of the universe. Over centuries and millennia, countless planets, moons, stars, nebulae and galaxies have been mapped, yet our inventory is by no means complete. There's always something new to discover.

In 1844, German astronomer Friedrich Bessel proved that it's possible to discover unknown celestial bodies without actually seeing them. He noticed deviations in the motions of the bright stars Sirius and Procyon. He could only properly explain these movements by hypothesizing that both stars have a small invisible companion star. Both stars do, in fact, have a companion, which was found later: a weak, compact white dwarf, whose gravitational force somewhat influences the movement of the main star.

Neptune's existence was similarly deduced, long before the planet was actually observed. In 1781, William Herschel discovered the planet Uranus at a great distance outside the orbit of Saturn. However, at the beginning of the 19th century, it was clear that Uranus didn't adhere to the predicted orbital movement. It was as if it was attracted to another, invisible, celestial body. Based on the orbital deviations that he measured, the French astronomer and mathematician Urbain Le Verrier calculated where this mischief-maker should be in the sky. In September 1846, this new planet – Neptune – was discovered at the Berlin Observatory, almost exactly at the precalculated position.

The companion stars of Sirius and Prokyon are visible with a telescope, as is the planet Neptune. However, it has become clear that even *invisible* objects can be detected through the gravitational pull they exert on their surroundings. It is precisely this approach that has convinced astronomers of the existence of dark matter in space. Precision measurements of the movements of stars and gas clouds in other galaxies played a decisive role. Still, no one has an exact idea of the true nature of this mysterious dark matter. What we do know is that there must be about five times as much of it as there is "normal" matter, in the form of atoms and molecules. Everything else is hidden – so far.

The first evidence of the existence of dark matter was found in the early 1930s by Dutch astronomer Jan Oort. Oort observed the speeds at which stars move "up" or "down" in relation to the central plane of the Milky Way. These speeds are influenced by the gravitational force of all matter in the Milky Way. In 1932, he deduced from the measured velocity distribution that there must be more matter in the vicinity of the Sun than one would apparently assume.

A year later, Swiss-American astronomer Fritz Zwicky came to similar conclusions about the amount of matter in galaxy clusters: they must also contain a lot of "dark matter." However, the most convincing evidence for the existence of dark matter came from measurements on other galaxies in the 1970s. The heavier a galaxy is, the faster the stars will move around the center – especially in the outer regions of the galaxy. If we can measure the rotational speed of a galaxy at different distances from its center,

A sparkling whirlpool

NGC 300 is a beautiful spiral galaxy located 6 million light-years from Earth, in the constellation Sculptor. You'd expect the rotational speed of a galaxy such as this to decrease further from the center. However, the outer regions show almost the same rotational speed as the inner ones – an indication of the presence of dark matter.

The halo in Leo
The impressive spiral galaxy M96 is located in the constellation Leo, 35 million light-years from Earth. Measurements of rotational speed have shown that it's enveloped in a vast, invisible halo of dark matter, just like other spiral galaxies. Without the stabilizing gravitational effect of such a halo, spiral galaxies could not exist.

then we can fairly accurately derive its total mass using these values. And if this mass is much larger than the mass of all visible stars, star clusters, gas nebulae and dust clouds combined, then we must assume that large quantities of dark matter exist.

But wait, how can these rotational speeds ever be determined? We aren't able to see how a distant galaxy spins in the course of a human life! From our point of view, everything seems to stand still due to the enormous distance. Even the stars in a nearby galaxy such as M31 or M33 don't travel more than a few dozen microarcseconds per year in our earthly sky; this minimal sideways motion is hardly measurable. However, astronomers have much less trouble measuring the radial motion of stars – toward us and away from us, along the line of sight. Radial velocities of stars can be determined with precise spectroscopic measurements. Just as the siren of a passing ambulance sounds slightly louder or softer depending on the speed at which the vehicle is moving toward or away from you, starlight shows a slight change in wavelength as the light source approaches or moves away from us. The greater the blue or red shift, the higher the radial velocity of the galaxy.

In 1970, American astronomers Vera Rubin and Kent Ford were the first to accurately determine the rotational speed of the Andromeda Galaxy in this way, up to about 22,000 light-years from the center. But the real breakthrough came eight years later, with the work of Dutch radio astronomer Albert Bosma. Bosma used Westerbork's brand-new radio telescopes – a total of 12 cooperating parabolic antennas, each 82 feet (25 m) in diameter – to measure the rotational velocities of hydrogen gas clouds on the outskirts of some 25 galaxies. These hydrogen clouds are at much greater distances from the center than the stars measured by Rubin and Ford, which helped to confirm their conclusions.

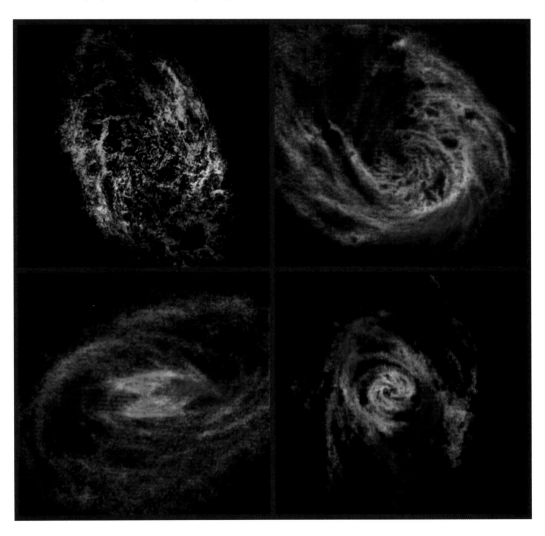

Lit-up spiral galaxies

The velocities of gas clouds in other galaxies can be measured using a radio telescope. Gas that moves toward us is depicted in blue; gas that moves away from us is depicted in red. Astronomers determine the gravitational distribution in the galaxy by measuring rotation patterns. Time and again, galaxies appear to contain a lot of invisible dark matter.

A dark secret

Situated in the constellation Ursa Major, M81 is also called Bode's Galaxy, after German astronomer Johann Elert Bode. Despite its distance of almost 12 million light-years from Earth, individual stars within the galaxy are visible in this Hubble mosaic. This galaxy must contain vast amounts of dark matter, but its true nature is unknown and has never been directly observed.

Today, measuring the rotational speeds of galaxies has become routine, with large radio observatories like the National Radio Astronomy Observatory's Very Large Array playing an important role. Time and again, something turns out to be incorrect. Due to the distribution of stars and gas clouds, one would expect that the rotation velocities would slowly decrease the further away you measure from the galaxy's core. In reality, however, it has been found that the rotational speeds in the outer regions of the galaxies are more or less constant, regardless of the distance from the center. This can only mean one thing: the galaxies are embedded in a giant dark matter halo, and the combined gravitational pull of all this dark matter gives the observable outer regions of the galaxy a high rotational speed.

But not everyone is convinced that dark matter exists, not least because physicists have never succeeded in capturing the mysterious particles in a laboratory on Earth, despite large-scale searches. Admittedly, the rotational properties of galaxies can't be determined from the calculated gravitational effect of visible matter. However, it's an assumption that our understanding of gravity is applicable to all circumstances, including the thin outer regions of galaxies. Perhaps gravitational force simply behaves differently there and we apply the wrong formulas, so it's no wonder that we arrive at inexplicable results.

A small group of doubters courageously resists the prevailing notion that the universe contains huge amounts of dark matter. With their "modified Newtonian dynamics" (MOND) theory, they can explain the observed velocity distributions in galaxies without dark matter simply by adapting the formulation of the theory of gravity. However, it seems to be proverbial tilting at windmills, as the existence of dark matter also derives from completely different observations, such as the distribution of galaxies in space, the properties of cosmic background radiation (the "echo" of the Big Bang), the gravitational lens effect of galaxies and galaxy clusters, and so on.

Many of these topics will be covered later in this book; here I've only touched on them to make it clear that the mystery of dark matter likely isn't easy to solve. What Shakespeare wrote about 400 years ago seems to be true: "There are more things in heaven and earth than are dreamt of in your philosophy."

In the end, research on galaxies defines our understanding of the universe. The dignified spinning spiral galaxies, the multiform barred spirals, the elliptical giants – they're all building blocks of the universe as well as springboards in our never-ending quest to decipher the mysteries of the cosmos.

Fast rotation

The Pinwheel Galaxy, also known as M83, is located in the southern constellation Hydra. Measurements taken with radio telescopes at a great distance from the center indicate the presence of clouds of cool hydrogen gas. The high rotational speeds of these clouds must be a consequence of the presence of large amounts of dark matter.

INTERMISSION

The Expanding Universe

Galaxies are unimaginably far away – millions or even billions of light-years. A light-year is the distance light travels at a speed of around 185,000 miles per second (300,000 km/s) over a period of one year: about 5.9 trillion miles (9.5 km). The Fornax Cluster, which is depicted here, is about 60 million light-years from Earth. Furthermore, distances between objects in the cosmos are constantly increasing as a result of the expanding universe. This cosmic expansion is revealed in the redshift of light from distant galaxies: during the long journey to Earth, the wavelength of the radiation emitted is stretched as the empty space itself expands. This redshift is therefore a measure of the time the light took to travel and is thus a reliable measure of distances in the observed galaxy.

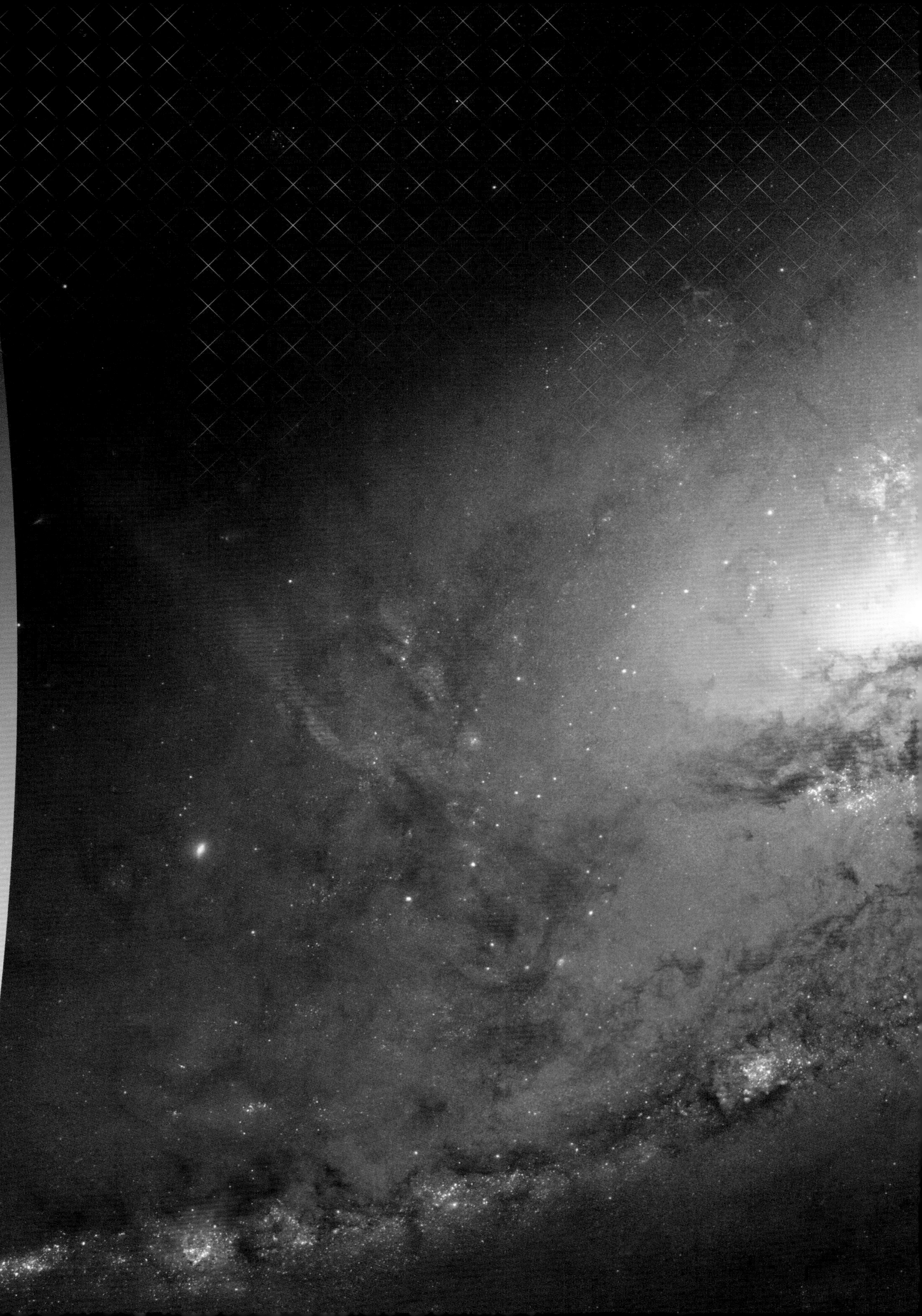

Additional arms
The spiral galaxy M106, about 20 million light-years from Earth, in the constellation Canes Venatici, is a so-called Seyfert galaxy with an active black hole in the center. Infrared observations (shown here in red) revealed two additional spiral arms bending upward and downward from the galactic plane. These spiral arms are likely the result of activity in the center.

Monsters and Gluttons

Dancing Galaxies

Humankind is a cosmic flash in the pan. *Homo sapiens* appeared and will disappear within the blink of the eye.

In the few hundred thousand years that we've inhabited Earth, the appearance of the universe has hardly changed. A human life is simply far too short to be aware of the evolution of the universe. Given the life span of the cosmos, a century is no more than a very short entry in the 14-volume encyclopedia of the universe.

We see the Moon going through its phases and the planets wandering through the constellations. Every now and then a comet appears in the nightly firmament, and a lucky few can occasionally observe the explosion of a distant star. However, the universe generally looks unchangeable to us. To us, the world of galaxies is the epitome of stability. If we could accelerate time, we would see spiral galaxies swirling around, dust clouds shrinking and star clusters lighting up. We could see dwarf galaxies and globular clusters buzzing around their mother galaxy like bees around a hive. And, ultimately, we could comprehend how quickly the galactic building blocks of the universe themselves change their form and appearance throughout cosmic history.

We may not think about it, but the question of "nature versus nurture" also plays a role in the universe: is everything preprogrammed by nature or shaped by the environment? Just as it is often difficult to understand whether character traits in humans are genetic or learned, we don't always know for certain the extent to which galaxies' traits are "innate" or

A wobbling disk

The central plane of the ESO 510-G13 galaxy is curved like the brim of a hat. Such distortions are usually caused by another galaxy's interfering gravitational forces, but in this case the source has not been identified. The galaxy is located 150 million light-years away from Earth, in the constellation Hydra.

A neighborhood dispute

Located 300 million light-years from Earth, in the constellation Andromeda, are these two asymmetric spiral galaxies, known together as Arp 273. As they passed each other quite closely, they were deformed by mutual tidal forces. The shock waves in the largest galaxy have triggered the formation of countless young blue stars.

A disrupted whirlpool

The Whirlpool Galaxy (M51) was the first galaxy identified as having a spiral structure. One of the spiral arms seems to have been pulled loose by the gravitational force of the small companion galaxy NGC 5195. These two galaxies are located 25 million light-years from Earth, in the small constellation Canes Venatici. The Whirlpool Galaxy can easily be seen with binoculars.

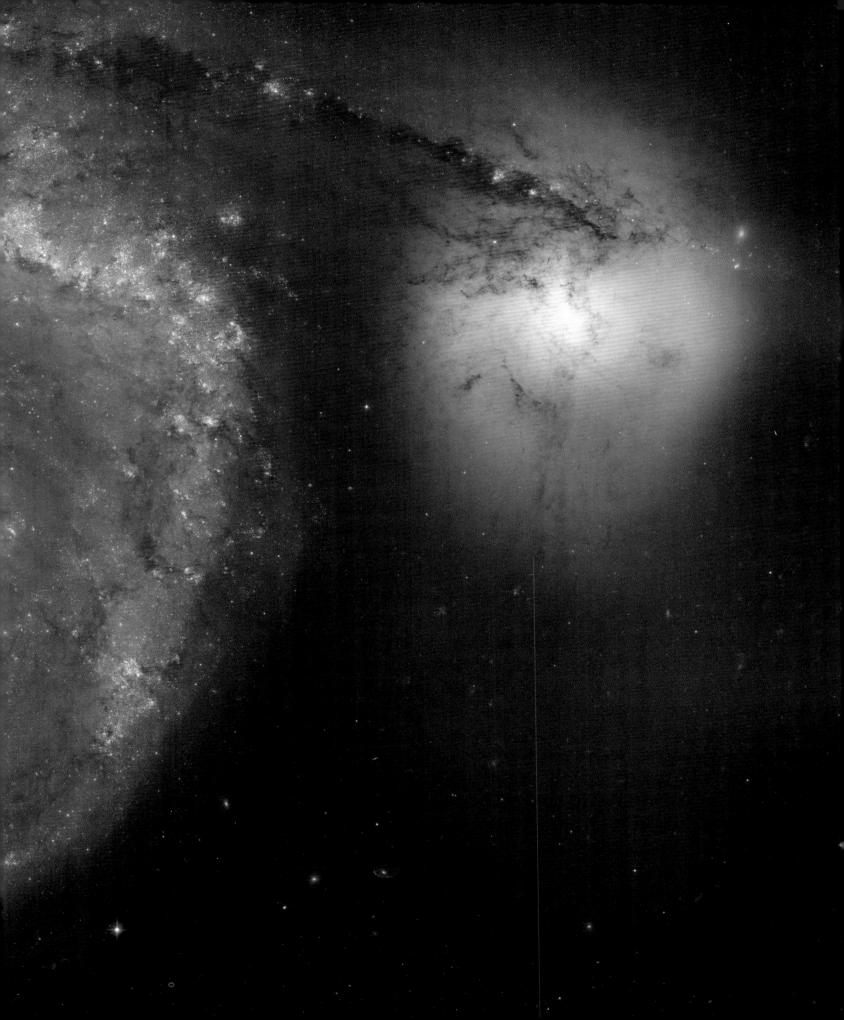

Appearances can be deceiving

Galaxies that appear close together in the sky are often located very far away from each other in relation to Earth. This is the case here: the "blue" galaxy is closer, causing the dust clouds in the disk to darken against the bright background of a distant galaxy (NGC 3314). There is therefore no gravitational interaction here.

"acquired." Until we're able to witness the individual lives of galaxies, we'll never get beyond speculation.

That eye-catching central bar – was it there from the beginning, or did it form later on? And if so, how exactly did it form and how long did it take? When it comes to the arms of a spiral galaxy being tightly or loosely wound, is this an unchangeable property of every single galaxy, or does it change over billions of years? Do the mysterious lenticular galaxies really represent an evolutionary transition between two types of galaxies, and how and in which direction does this evolution take place?

One thing's for sure: just like humans, galaxies are usually not isolated phenomena. A hundred years ago, astronomers like Edwin Hubble spoke of "island universes" – that all galaxies, apart from a few exceptions, were part of a larger whole, like a "social structure," if you will. The Milky Way, the Andromeda Galaxy, and the Triangulum Galaxy, along with dozens of dwarf galaxies, form the so-called Local Group; elsewhere, in the universe, we encounter gigantic clusters of galaxies and superclusters (which are the subject of the next part of this book). And just as every human is influenced by their environment, every galaxy is susceptible to influences from its closest neighbors.

We aren't able to see exactly how galaxies influence each other over the course of cosmic history. Fleeting ephemera such as ourselves must be satisfied with a momentary snapshot – similar to how people may speculate about inter-personal relationships when looking at old family photos. Fortunately, the interaction between two galaxies is a much simpler and more predictable undertaking than the complex

A group process

Stephan's Quintet is a group of five galaxies located 300 million light-years from Earth, in the constellation Pegasus. Two of these galaxies, which can be seen slightly to the right of the center of this photo, are almost fused together. Stephan's Quintet has been deformed by mutual tidal forces. The exception is the blue galaxy at the top left. It seems to be a small galaxy in the foreground, 40 million light-years away.

social-psychological relationships among humans. The universe is dominated by the influence of gravity, which can be accurately described with a handful of simple and reliable formulas.

The fact that each galaxy's mutual gravitational pull exerts its influence on others was revealed by Lord Rosse's observations in the 19th century. His pencil sketches of M51 (a galaxy located in the constellation Canes Venatici) first revealed the impressive spiral structure that gave the fleck of light its nickname, the "Whirlpool Nebula." Lord Rosse's drawings also clearly showed that one of the spiral arms seemed to be slightly distorted: it was pointing in the direction of a smaller neighboring nebula. Could it be possible that the original symmetrical shape of the large galaxy was disturbed by the gravitational pull of the smaller companion?

Since Lord Rosse's time, the Whirlpool Galaxy – located 25 million light-years away from Earth – has been observed in great depth by the Hubble Space Telescope, among others. Today, there's no doubt that the asymmetric spiral structure is indeed caused by the smaller galaxy NGC 5195. The smaller galaxy itself is also enormously deformed, so much so that astronomers don't venture to determine its type with certainty. As seen from Earth, it's clear that NGC 5195 is located slightly behind the Whirlpool Galaxy; the dust present in the distorted spiral arm of M51 is clearly visible in the silhouette against the stellar glow of the smaller galaxy.

Scientists have also discovered galaxies that pass quite closely to each other in other parts of the universe. For dozens of millions of years – just the blink of an eye in the universe's calendar – the galaxies continue to mutually influence each other, and the consequences of this influence can still be visible a few hundred million years later. We can compare this with the way a brief, intensive meeting with a charismatic and inspiring person can give our life a new direction over a long period of time.

The gravitational pull of a passing companion can cause a fine curvature in the plane of a spiral galaxy. Spiral arms aren't just deformed and elongated, they're often pulled out of symmetry. In some cases, a long bridge of thin hydrogen gas, littered with sparkling stars, forms between the two passing galaxies. And when compressions and shock waves occur in the extracted gas, new stars are created.

The key term to describe these interactions is tidal effect. In our daily lives, we understand tidal effect primarily through the ebb and flow of currents, which are caused by the force the Moon exerts on the Earth's oceans.

The tidal effect occurs when two relatively large objects based on the distance that separates them are influenced by gravity. The thin layer of water on the Earth's surface is a good example of this: this layer of water has a diameter of about 7,800 miles (12,500 km), which is about three percent of the distance between Earth and the Moon. This means that the gravitational force of the Moon is noticeably stronger on the side of Earth that is facing the Moon than on the side that is not facing the Moon. We understand this difference in gravitational force as the tidal force of the Moon.

The same applies to interacting galaxies. The motion of the two galaxies – the way in which their orbits are deflected – is determined by mutual gravitational force. However, each galaxy experiences a stronger gravitational effect on one side and, therefore, a stronger deflection on the other. As a result, the galaxies are slightly stretched along their imaginary connecting line. (In the same way, the Earth's malleable "water bowl" is easily attracted by the tidal forces of the Moon, which is why there are always two tidal crests: one more or less "beneath" the Moon and another on the opposite side of Earth.)

The effects of the tidal interaction between two close galaxies depends not only on their mass, speed and distance, but also on the orientation and internal structure of the two systems. Much more complicated is the interaction between three or more galaxies, e.g., in the case of Stephan's Quintet, a compact group of galaxies hundreds of millions of light-years from Earth.

The clash of galaxies often results in collisions and mergers. What starts as subtle tidal distortion quickly turns into a chaotic scene of matter rotating against matter, long spiral arms, ejected gas and dust streaks, along with the birth waves of new stars. In the next chapter we'll learn more about these spectacular cosmic traffic accidents.

Even our own Milky Way hasn't entirely escaped the consequences of tidal effect. The Large and Small Magellanic Clouds, located diagonally below the Milky Way, so to speak, cause a slight curvature in the central plane of the galaxy. In the distant future, our home galaxy will be deformed even more by the gravitational influence of the Andromeda Galaxy when these two spirals approach each other in a head-on collision.

A distorted spiral

Spiral galaxy NGC 2442 is located 55 million light-years from Earth, in the constellation Volans. Due to its unusual shape, it's also known as the Meathook Galaxy. One of the spiral arms is very distorted and contains many brightly shining regions where stars are being formed. This distortion is almost certainly the result of another galaxy that passed closely by it some time in the past.

Collisions and Mergers

The motion picture *Frida* (2002), by American director Julie Taymor, features a spectacular slow-motion scene of the dramatic bus accident that occurred on September 17, 1925, in which Mexican artist Frida Kahlo was severely injured at the age of 18. Time is slowed down so much that the short, fatal blow is almost transformed into a graceful ballet of falling bodies, distraught looks, bent steel and flying glass fragments. Any slower and the clock would stop – time freezes and as a spectator you're caught in a single moment.

I often think of this scene when looking at the spectacular Hubble photo of the Antenna Galaxy. Here we have two galaxies that collided head-on and that owe their nickname to the long, curved streaks of gas and stars that had been drawn even before the collision occurred by mutual tidal forces. Here, too, the spectator becomes witness to a still photograph from a disaster film. Based simply on this snapshot,

A cosmic antennae

When these two galaxies – NGC 4038 and NGC 4039 – collided, long tails of gas and stars were hurled outward by mutual tidal forces. These galaxies are located 45 million light-years from Earth, in the small constellation Corvus. They were discovered as early as 1785 by William Herschel. In a few hundred million years, they will have merged together.

A second youth

The collision of the Antennae Galaxies triggered a birth wave of new stars in each galaxy. This Hubble close-up shows glowing gas nebulae, bright young star clusters and ejected dust clouds. The future collision of our Milky Way with the Andromeda Galaxy will also trigger such large-scale star-forming activity.

Wheel of fortune

The Cartwheel Galaxy (ESO 350-40), located 500 million light-years from Earth, owes its remarkable shape to a frontal collision with a smaller galaxy – presumably the deformed blue spiral galaxy at the top left. This smaller galaxy flew across the larger one, and shock waves blew gas and dust into a gigantic ring in which new stars are forming.

it's not easy to reconstruct the actual course of events of this cosmic traffic accident, which took place at a distance of tens of millions of light-years from Earth.

The two galaxies, located in the relatively small constellation Corvus, were discovered in 1785 by William Herschel. They were given the designations NGC 4038 and NGC 4039. In the 1960s, American astronomer Halton Arp included them together as number 244 in his *Atlas of Peculiar Galaxies*. Everything seemed to suggest a galactic collision. But how exactly does a collision such as this happen? And why these asymmetric tidal tails?

Previously, I explained that the interaction between two neighboring galaxies is almost entirely determined by their gravitational pull, which can be accurately determined using mathematical equations. However, there's no single formula with which we can calculate the net result of a cosmic collision, even if the size, structure, orientation and speed of the two hapless galaxies are fully known. Only by dividing the process into numerous small time steps can we understand the true course of events. We have to calculate over and over how the motion of each individual star is determined by the collective gravitational force of all the other stars in the two galaxies.

Today's supercomputers can crack this kind of hard nut in just minutes, but in the early 1970s, it took the simple calculators of brothers Alar and Juri Toomre many weeks to produce a reasonably acceptable result based on the calculations of a few hundred small test parts. I myself remember struggling in the early 1980s with the frustration of programming my Commodore 64 home computer to simulate the collision of two galaxies – it took countless hours and the results were not very good.

Nevertheless, the Toomre brothers were the first to plausibly explain that the tails of NGC 4038 and NGC 4039 were formed by tidal forces. And thanks to advanced computer simulations, we now have a pretty good understanding of how the collision of these antenna galaxies occurred and how it will continue. It's been proven that they collided about 600 million years ago, when they passed through each other. About 300 million years ago, the long tidal tails formed. Since then, they've slowed down too much to escape the mutual influence of gravity;

Mixed doubles

When two colliding galaxies have slowed down enough, they eventually merge into one large galaxy. For NGC 2623, situated 300 million light-years from Earth, in the constellation Cancer, that time has come. Stars are still being born in the long tidal trails, but almost nothing of the original spiral structure of the two galaxies can be seen today.

An exploding cigar

This photomontage of the Cigar Galaxy (M82), located 13 million light-years from Earth, in the constellation Ursa Major, shows a great deal of star-formation activity and signs of explosive phenomena in the center, probably due to the galaxy's interaction with the neighboring spiral galaxy M81. We can also see visible light (orange, yellow and green), X-ray radiation (blue) and infrared radiation (red).

in about 400 million years, the two galaxies will merge to form a giant elliptical galaxy. The exact same fate awaits our own Milky Way and the neighboring Andromeda Galaxy in a few billion years. It may sound incredible that two gigantic galaxies can pass through each other, but galaxies aren't solid structures like cars or buses. Stars within a galaxy are relatively far away from each other: the distance between the Sun and the closest star, Proxima Centauri, is about 25 trillion miles (40 trillion km), and the diameter of the Sun is just over 900,000 miles (1.5 million km). That means the distance between the two stars is about 25 million times as large as the diameter of the Sun. In other words, galaxies consist primarily of empty space.

Of course, when two galaxies approach each other and collide, the movements of the stars within these galaxies are affected by the gravitational force of all the other stars. However, due to the large distances between these stars, the probability of individual stars actually colliding is effectively zero. From this point of view, two galaxies pass through each other as easily as do two swarms of mosquitoes; in reality, it is easier still.

The two galaxies are, of course, slowed down by the opposing gravitational pull, which in most cases ends in a merger. But there are other factors involved: the space between the stars isn't really empty, and thin gas and fine dust in one galaxy immediately collides with the interstellar matter in the other galaxy as the collision initially happens. Both galaxies then become much denser, unleashing slow-moving shock waves, which spread out in all directions. In unexpected places, new collections of young, hot stars emerge, some of which will explode again as supernovas in just a few million years. Thus, a relatively quiet and tranquil galaxy (like our Milky Way) can turn into a tumultuous arena of glowing gas, brightly shining stars and dazzling explosions following a cosmic collision.

The birth waves of new stars occur not only *within* the two clashing galaxies, but also in the tidal tails. They also contain large quantities of gas and dust, from which complete star clusters are born once again. Regions of brightly illuminated star formations and sparkling groups of new stars thus mark the areas where the collision resonates most strongly and where the sloshing gas temporarily reaches the highest density, just as foam crests crown the highest ocean waves.

How the density waves expand depends on the angle at which the two galaxies collide. The widely known Cartwheel Galaxy is the result of a collision in which a relatively small galaxy crashed almost directly from above onto a spiral galaxy, similar to a stone falling into a pond. The impact caused large quantities of gas to be blown outward at high speed. Shock waves from this gas led to the formation of new stars in a remarkably bright ring at a great distance outside the spiral galaxy. Radially aligned tidal tails give the galaxy a distinctive appearance, to which it owes its nickname.

Even after the turmoil of the collision has subsided, traces of the disaster can often still be seen with a little effort. When a large galaxy devours a smaller galaxy, not much remains of the original symmetrical structure. Once a large elliptical galaxy is formed, it's often surrounded by concentric shells of hot gas and weak stars; all matter is blown outward in the reverberations of the cosmic collision and is only visible in long-exposure photos. Astronomers often discover unusual movements in the core of the newly formed galaxy: stars that circle around the center against the normal direction of rotation.

Although serious traffic accidents happen every day here on Earth and can drastically change the existence of those involved – as Frida Kahlo's tormented life shows – they have never played a decisive role in the evolution of humanity as a whole. This is not, however, the case in the universe. Nearly all galaxies will eventually collide with a small or large companion galaxy and be affected by this cosmic collision. In fact, most galaxies owe their current appearance to earlier episodes of galactic cannibalism, as we will see in the last part of this book.

Galaxies are the building blocks of the universe. However, these blocks are not static and cannot be changed like Lego bricks. They were created from smaller precursors, influence each other, change their shape, collide with each other and merge to form larger structures. Slowly but surely, it's becoming increasingly clear how this galactic evolution has taken place over billions of years.

Post-traumatic stress

The consequences of a cosmic catastrophe that unfolded when two galaxies collided about a billion years ago are still visible: thin shells of gas and dust and a small central spiral of stars rotating in the "wrong" direction. This galaxy, NGC 7252, is about 200 million light-years from Earth, in the constellation Aquarius.

Active Cores and Quasars

The active center

The center of the barred spiral galaxy M77, situated 47 million light-years from Earth, in the constellation Cetus, contains a huge black hole. M77 is the first active-nucleus Seyfert galaxy that scientists discovered. This photo, taken by the European Southern Observatory's Very Large Telescope, also shows the gas clouds being ejected into the weak outer regions of the galaxy.

Looking at the detailed photos in this book, it's hard to imagine that just 100 years ago little was known about galaxies.

As larger and larger telescopes were built in the 18th and 19th centuries, the number of known cosmic nebulae increased rapidly: from the 100 or so that were cataloged by Charles Messier to the almost 8,000 listed in John Dreyer's New General Catalogue. But as far as their nature, location and dimensions were concerned, astronomers were in the dark.

It wasn't until the early 1920s that Edwin Hubble determined the distance from Earth to the Andromeda Galaxy. As a result, it could no longer be denied that numerous spiral nebulae in the starry sky – just like their elliptical counterparts, as it was later discovered – were at a great distance outside our own Milky Way. However, the fact that many of these nebulae possess very special properties was known much earlier. Even if you don't know anything about the distance or the true nature of a nebulous object, you can accurately measure the light it emits.

Astronomers have been using the technique of spectroscopy for many years. During this process, captured light is dispersed into spectral colors, the same way that raindrops disperse white light from the Sun. This allows us to determine very precisely how the energy of the emitted light is distributed in the different wavelengths. And since each chemical element absorbs or emits radiation at specific wavelengths, these spectroscopic measurements reveal information about the chemical composition of a celestial body. It's no exaggeration to say that spectroscopy – along with the invention of the telescope – was the greatest technological breakthrough in the history of astronomy.

At the beginning of the 20th century, numerous nebulae had already been detected using absorption lines, which are wavelengths of light that are absorbed by relatively cool gas atoms, leaving dark lines visible in the spectrum of the nebula. But in 1908, astronomers made a surprising discovery: the light emitted by the M77 nebula (located in the constellation Cetus) showed bright emission lines instead of dark absorption lines. This suggested that the (also remarkably bright) nucleus of the galaxy contained extremely hot gas that emitted radiation in particular wavelengths. Once it became clear that spiral nebulae were actually galaxies outside of the Milky Way, astronomers had to find an explanation as to why some of these galaxies appeared to have a much more "active" nucleus than others. In the 1950s, American astronomer Carl Seyfert published new measurements of six of these extraordinary systems, including M77. (These galaxies are now known as Seyfert galaxies, and it is clear that about one in ten galaxies belongs to this category, even if the emission lines aren't always the same.) However, in the middle of the last century, scientists did not know why the gas atoms in the centers of these galaxies were in such a turbulent state.

The fact that strange scenes often take place inside the nuclei of galaxies became apparent as early as 1918, when American astronomer Heber Curtis – the most important proponent of the theory that spiral nebulae are extragalactic – took a detailed photo of the M87 nebula, located in the constellation Virgo. Although M87 did not appear to have spiral structure, the photo showed a bizarre, straight-as-an-arrow beam of light emanating from the center of the circular nebula, as though a narrow band of

A ring of fire

A luminous ring of newborn stars, 5,000 light-years in diameter, marks the center of the Seyfert galaxy NGC 1097, which hides a massive black hole. Narrow dust trails can be seen in the silhouette against the bright background of the galaxy. NGC 1097 is located in the constellation Fornax (chemical furnace), which is 65 million light-years from Earth.

energy was being catapulted outward from the center.

Shortly after the Second World War, astronomers measuring cosmic radio waves discovered that the constellation Virgo contained a powerful source of radio radiation that corresponded to the position of M87 in the sky. The constellations Cygnus and Centaurus also seemed to have brighter radio sources coinciding with a large elliptical galaxy. Virgo A, Cygnus A and Centaurus A, as the three bright radio sources are officially called, are among the next – and thus brightest in terms of Hubble's classification – examples of a completely new class of galaxy: the radio galaxies.

Thanks to the increasing sensitivity of large radio telescopes, many thousands of radio galaxies are now known, most of which are located very far from each other and from Earth. Like Virgo A (M87), they all feature prominent jets (called beam currents) that seem to originate from the center of the galaxy. In most of these galaxies, two of these jets are being blown into space in opposite directions. The

A black heart

M87 is a galaxy of superlatives. It contains several trillion stars and thousands of globular clusters. At its center is a gigantic black hole that is about 6 billion times heavier than the Sun. This Hubble photo clearly shows the long jet caused by the black hole. M87 marks the center of the Virgo Cluster.

Big mouth

3C 348 is the main galaxy in the Hercules Cluster. The huge black hole at the center of this elliptical galaxy is 2.5 billion times heavier than the Sun. It blows two jet streams of charged particles into space in opposite directions. The Very Large Array radio telescope was used to record these jets, which were 1.5 million light-years long (shown here in pink).

The first quasar

At the end of 1962, Maarten Schmidt discovered that the mysterious radio star 3C 273 wasn't a normal star in the Milky Way, but rather a quasar, which is the active nucleus of a distant galaxy located about 2 billion light-years from Earth. This dazzlingly bright core outshines the rest of the galaxy, but the jets of the quasar can still be seen in this Hubble photo.

observed radio radiation originates from very fast moving electrons within these energetic currents and especially from large radio "clubs" at the tips, where the jets come into contact with (and are slowed down by) thin intergalactic matter.

At the beginning of the 1950s, however, there were very few known sources of radio radiation in the sky. Radio astronomy would experience its peak during the 1950s. The radio observatory in Cambridge, England, published several catalogs of radio sources, the most detailed of which was the Third Cambridge Catalogue of Radio Sources (3C), which appeared in 1959 and contained several hundred objects. In many cases, such radio sources coincided with a galaxy. However, 3C also catalogued sources that couldn't be traced to a single known object, primarily because their exact celestial position was not known. One of these sources was 3C 273, which, like Virgo A, is located in the constellation Virgo.

In the fall of 1962, the Moon passed in front of this radio source, allowing for a much more accurate determination of its position in the starry sky. The relatively bright source of radio radiation seemed to coincide with a completely inconspicuous star, so it was assumed that it must be a star in our own Milky Way. However, when Dutch-American astronomer Maarten Schmidt identified the spectrum of this mysterious radio star at the end of 1962, he discovered that it was actually an object about 2 billion light-years away. Therefore, 3C 273 had to be a very distant – and remarkably energetic – radio galaxy. It didn't take long for more of these quasi-stellar radio sources, as they were known at the time, to be found; today, they are called quasars.

Seyfert galaxies, radio galaxies and quasars are examples of what are generally referred to as active galaxies. In all cases, these galaxies have a high-energy nucleus (referred to as an AGN, which is short for active galactic nucleus) and are often surrounded by a region where bright emission lines are generated

A dusty center

The central part of the active galaxy NGC 5128 (Centaurus A) offers a chaotic picture of bizarre dust veils and regions of bright star formation. The center of the galaxy (top left) is dominated by a colossal black hole that weighs about 55 million times more than the Sun. In the immediate vicinity of this black hole, bundles of electrically charged particles are blown into space (not visible in this Hubble photo).

by hot gas. Active galaxies emit not only large amounts of visible light and radio radiation but also high-energy ultraviolet and X-ray radiation. The elliptical galaxy M87, for example, isn't just known as a bright radio source (Virgo A) but also as a powerful source of X-rays (Virgo X-1), which was first discovered during a rocket experiment in 1965.

Based on their observed properties, the different types of active galaxies have been further subdivided into subclasses: type I and II Seyfert galaxies; FR class I and II radio galaxies (the letters stand for Fanaroff and Riley, the two astronomers who introduced this subdivision); and "radio loud" and "radio silent" quasars. Astronomers have also identified a number of other types of AGNs, including blazars, BL Lacertae objects (named after the prototype in the lizard constellation, Lacerta) and optically violent variable (OVV) quasars.

Just as Edwin Hubble divided galaxies into spiral, barred spiral and elliptical, including a number of subclasses, astronomers have developed an impressive classification system for active galaxies dating back to the middle of the last century. It shows that scientists are always eager to record similarities and differences, as this often leads to a better understanding of the underlying causes of this diversity.

In the case of the active galaxies, the external differences between the individual types are most likely due to a difference in orientation. The angle at which a galaxy is viewed affects how scientists are likely to categorize it. For example, when a galaxy is viewed directly from above, we are looking directly into the high-energy jets. When a galaxy is viewed from the side, however, the jets are more clearly visible, but this more or less obscures the extremely bright core. Within this so-called unification model, all active galaxies have one characteristic in common: the driving force in all of these high-energy galaxies is a supermassive black hole, just like the one hidden at the center of our own Milky Way.

Supermassive Black Holes

It's estimated that our universe contains more than a hundred billion galaxies. Nearly every one of these systems has a dark secret deep within it: a large and voracious black hole. Some of these cosmic gluttons hold back, while others reveal their presence by devouring gas clouds and even entire stars in their vicinity. The most extreme specimens manifest themselves as glowing beacons of high-energy X-ray radiation. These supermassive black holes are as puzzling as they are numerous – their origins are still a mystery.

The existence of black holes was predicted by Einstein's theory of relativity. When a heavy star dies in a catastrophic supernova explosion, the nucleus collapses into a compact neutron star or, if the star is heavy enough, into a black hole, which is a mysterious object that has such a strong gravitational field that even light cannot escape it. For dozens of years, black holes were considered theoretical curiosities. However, in 1971, astronomers discovered that the Cygnus X-1 X-ray source actually housed such a bizarre object, and it was 15 times heavier than the Sun. That same year, British cosmologists Donald Lynden-Bell and Martin Rees suggested that the center of the Milky Way might hide an even larger black hole. As I described earlier in this book, the existence of this supermassive black hole (with a mass of about 4 million Suns) has been conclusively proven, and the Milky Way is certainly not the exception. It is generally assumed that nearly every galaxy in the universe contains a similar supermassive black hole.

Black holes themselves are, by definition, invisible. Their presence is revealed through their strong gravitational field. Based on measurements of the velocities of stars in M32, one of the elliptical companions of the Andromeda Galaxy, it was deduced in 1987 that the core of that galaxy must contain a black hole several million times heavier than the Sun. The center of the Andromeda Galaxy also contains a black hole that's at least 100 million times heavier than the Sun.

But even when a galaxy's too far away to measure the velocities of individual stars, a central black hole can still be detected. Absorbed gas will accumulate in a flattened, rotating disk before it disappears over the event horizon of the black hole. The gas in this accretion disk, as it's called, becomes so hot that it emits X-rays. The total amount of X-ray radiation is a reliable measure of the mass of the black hole. In addition, a small part of the absorbed matter is blown into space at enormous speeds in opposite directions along the axis of rotation of the accretion disk. As shown earlier, virtually all active galaxies have these jets; they're likely to be accelerated by strong magnetic fields in the immediate vicinity of the central black hole. There's little doubt that every radio galaxy has a supermassive black hole and that the enormous energy production of quasars is due to a giant black hole in the nucleus of the parent galaxy. In recent decades, astronomers have succeeded in actually "weighing" numerous supermassive black holes. The fearsome monster Sagittarius A*, at the heart of our own Milky Way, seems to be an extremely modest specimen of about 4 million Suns. A few hundred million times the weight of the Sun seems to be the rule rather than the exception in the universe, and many examples of even heavier black holes are known.

The record holder in a nearby universe is undoubtedly the black hole at the center of M87 – the elliptical galaxy within the constellation Virgo, also

An absorbing effect

Cygnus X-1 was the first black hole to be identified with certainty. In this image, we can see that it is absorbing the matter of an accompanying star. Before the gas disappears forever, it emits high-energy X-rays. The Milky Way probably contains several million stellar black holes such as Cygnus X-1, which formed after a supernova explosion.

A fossil giant

At first glance, you probably wouldn't expect it, but the elongated galaxy NGC 1277, located in the Perseus Cluster, contains a black hole that's probably several billion times heavier than the Sun. The galaxy consists almost exclusively of extremely old stars; it's a galactic fossil from the earliest days of the universe. NGC 1277 is 220 million light-years from Earth.

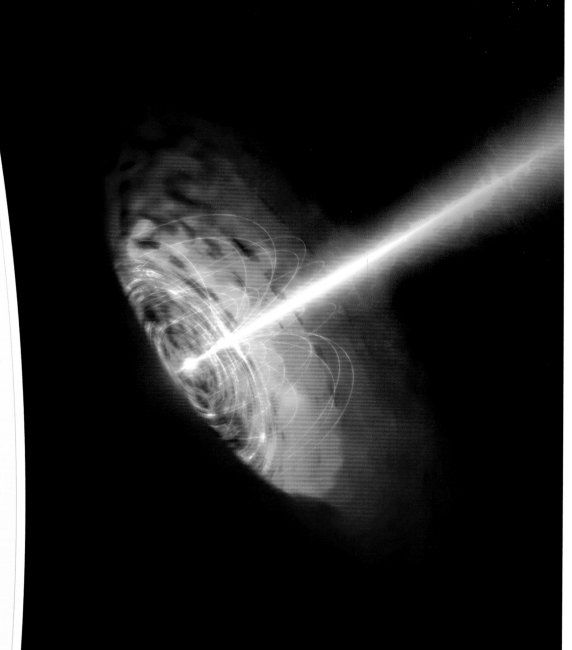

A close-up of a quasar

A supermassive black hole in the nucleus of a galaxy blows a high-energy jet of electrically charged particles into space, presumably under the influence of strong magnetic fields. The black hole is surrounded by a hot accretion disk; further away, there's a thick, almost opaque torus of cosmic dust.

known as the Virgo A radio source. Statistical velocity measurements of stars within the galaxy's core show that the central black hole must be 6.4 billion times heavier than the Sun – about 1,000 times heavier than the black hole in the center of the Milky Way. Although the distance between Earth and M87 (53.5 million light-years) is 2,000 times greater than the distance between Earth and Sagittarius A*, a group of scientists focused the Event Horizon Telescope on M87 in the hopes of capturing the world's first image of a black hole. they succeeded in April 2019, and the image quickly went viral around the world.

The black hole in the distant galaxy TON 618, 10.4 billion light-years from Earth, is 10 times heavier than Sagittarius A*. This suggests we're dealing with an "ultra-heavy" black hole, estimated to be 66 billion times heavier than the Sun. And there are still more massive black holes to be discovered.

There also seems to be a relationship between the mass of a galaxy (or its central bulge) and the mass of the supermassive black hole at its center. There are, of course, plenty of exceptions – the black hole in the Milky Way's nucleus is relatively light, and the black hole in the Andromeda Galaxy is one of the heaviest. In short, we can say that a galaxy five times as massive as Earth will likely contain a black hole five times as massive. This leads to the assumption that the growth of supermassive black holes is in some way linked to the growth of the parent galaxy.

By now it's clear that the presence of a heavy, active black hole can greatly influence the evolution

A monster in the Milky Way

Images of the hottest objects and structures at the center of the Milky Way were captured using X-ray telescopes in space. The numerous X-ray "stars" are stellar black holes and neutron stars. Sagittarius A*, an enormous black hole that weighs around 4 million times more than the Sun, is located in the bright center of the Milky Way. Its striking blue light is visible at the bottom right of this image.

A former record holder

With a little imagination, we can visualize what the distant quasar ULAS J1120+0641 looks like up close. The hot accretion disk around the central black hole is distorted by magnetic fields, while gas is shot away at high speeds along the axis of rotation. About 800 million years after the Big Bang, this quasar was one of the brightest objects in the universe.

A distant heavy-weight

The red dot at the center of this photo is ULAS J1120+0641, one of the most distant quasars ever discovered, situated 13 billion light-years from Earth, in the constellation Leo. Exactly in the bright nucleus of this galaxy is a black hole that weighs 2 billion Suns. We see the distant quasar as it looked when the universe was about 800 million years "young."

of a galaxy. The high-energy radiation from the accretion disk blows hot gas out through the galaxy, preventing the formation of new stars – from cooler gas and dust clouds – or even bringing them to a complete standstill. Conversely, the supply of large quantities of gas into the nucleus of the galaxy (e.g., after a collision with another galaxy) will lead to a substantial increase in the mass of the central black hole. When two galaxies collide and both have a black hole at their center, the two black holes will eventually merge into an extra-large and heavy specimen.

Supermassive black holes located in the center of galaxies have recently revealed a great deal of their secrets, yet astronomers still haven't solved two questions: how did the voracious monsters come into being, and how could they ever become so incredibly heavy? Ultimately, there are limits to the speed at which a black hole can grow. If too much gas flows in all at once, it is heated up to such an extent that the generated radiation creates counterpressure.

Today, the universe is about 13.8 billion years old. One might therefore assume that black holes have had plenty of time to grow to their present size. This presumably also applies to the black hole that is 6.4 billion times heavier than the Sun in the nucleus of M87. However, other black holes already had an incredibly large mass when the universe was still quite young. In the constellation Boötes (the herdsman), for example, a bright quasar was discovered 13.1 billion light-years from Earth. This means that the light of the quasar took 13.1 billion years to reach Earth. We therefore see the quasar as it looked 13.1 billion years ago, when the universe was only 700 million years old. Remarkably, however, the black hole at the center of this quasar has a mass of about 800 million Suns.

Another big question is: what did the very beginnings of the formation of supermassive black holes look like? During a normal supernova explosion, a black hole of at most a few dozen solar masses is formed. But the very first stars in the newly born universe were probably much heavier than the heaviest stars in the universe today. Perhaps these black holes left behind a few hundred solar masses. It's even possible that massive clouds of hydrogen gas collapsed under their own weight into black holes in this early prehistoric period. But even then there must have been an incredibly high growth rate during the earliest stages of the universe. Clearly, a straightforward solution to this riddle isn't likely. In the future, telescopes will be able to penetrate further into the universe and thus look back further into time. Perhaps in this way we'll be able to trace the earliest precursors of today's supermassive black holes in the very first galaxies. Once again, galaxies prove to be the key to unraveling the universe.

INTERMISSION

Big Eyes

Astronomers rely on large telescopes – both here on Earth and in space – to explore the universe. The James Webb Space Telescope (the successor to Hubble) is due to be launched in 2021 and has a mirror diameter of 21 feet (6.5 m). The Extremely Large Telescope of the European Southern Observatory, shown here, will be much larger. It's currently under construction on top of the Cerro Armazones mountain in northern Chile and will receive a segmented primary mirror with a total diameter of 128½ feet (39.2 m). Once completed, it will be by far the largest telescope in the history of astronomy. This new generation of telescopes allows astronomers to study galaxies at the edge of the universe and at the beginning of time. In addition, new observation instruments in other wavelength ranges, such as X-rays, microwave and radio waves, are constantly being constructed.

A packed furnace

Hundreds of individual galaxies can be seen in this photo of the Fornax Cluster taken with the VLT Survey Telescope in Chile. The galaxy cluster is located about 60 million light-years from Earth, in the constellation Fornax (chemical furnace). The spots of light around the brightest objects are the result of reflections in the telescope and camera optics.

Galaxy Clusters

Cosmic Collections

Look in the direction of the constellations Virgo and Leo on a clear spring night, and you'll see straight into the heart of a gigantic swarm of galaxies. To the right is the star Denebola (the Arabic name for the "tail of the lion"). It's a young, hot star located 36 light-years away. On the left is Epsilon Virginis (also called Vindemiatrix) in the constellation Virgo. It looks somewhat weaker than Denebola, but that's because it's 110 light-years further away; in reality, it is a luminous giant star.

In the seemingly starless region between these two stars is the Virgo Cluster, which is a gigantic cluster of galaxies tens of millions of light-years from Earth. They can't be seen with the naked eye, but a small amateur telescope is enough to observe them.

At the end of the 18th century, French astronomer Charles Messier discovered that an enormous concentration of nebulae was found in this part of the starry sky; one-sixth of the objects in his catalog are grouped here. In addition to these 16 relatively bright Messier objects, the galaxy cluster contains many hundreds of weaker galaxies. It is estimated that a total of about 1,500 galaxies are distributed over an area with a diameter of about 10 million light-years.

Astronomers currently have a fairly good picture of the spatial structure of the Virgo Cluster. It seems to consist of three partial clusters, each grouped around a large elliptical galaxy; one of these three is the giant galaxy M87, which contains a black hole weighing several billion Suns. Elliptical and lenticular galaxies are mainly found in the central part of the galaxy cluster, while spiral galaxies are mainly found at the edges of the cluster. Distance measurements have shown that the galaxy cluster doesn't have a spherical shape but extends relatively far in the direction of view.

The Virgo Cluster is not the only galaxy cluster, but with a distance of 54 light-years from Earth, it's probably the closest one to us. A little further north in the sky, in the inconspicuous constellation Coma Berenices (Berenice's hair) we find the Coma Cluster, which contains about 1,000 galaxies. The galaxies in the Coma Cluster are considerably weaker and appear much smaller in the sky since, at 320 million light-years, it's much further from Earth. Other known galaxy clusters are the Hercules Cluster, the Perseus Cluster, the Fornax Cluster, the Hydra Cluster and the Centaurus Cluster – all named after the constellation in which they're located.

American astronomer George Abell was the first to systematically explore galaxies in the late 1950s. As part of his doctoral research at the California Institute of Technology in Pasadena, Abell examined the photographic plates of the starry sky taken with the 4-foot (1.2 m) Schmidt telescope at the Palomar Observatory in southern California. The Palomar Observatory Sky Survey, which was partly financed by the American National Geographic Society, consists of nearly 2,000 glass plates on which many thousands of stars and galaxies are depicted. Armed with a lightbox, a magnifying glass and an incredible amount of patience, Abell went in search of clusters of galaxies. His first catalog, published in 1958, contained 2,712 clusters. When the list was later extended to include galaxy clusters in the southern starry sky that had been photographed at an observatory in Australia, the number increased to 4,073. The Coma Cluster is also known as Abell 1656; the Fornax Cluster is Abell S373. (Strangely enough, the large nearby Virgo Cluster doesn't bear an Abell designation – it's so large in the sky that it covers several photographic plates.)

The Virgo swarm

The Virgo Cluster, located in the constellation Virgo, is the closest large galaxy cluster to Earth. The center of the cluster is 54 million light-years away from us. In this overview photo, the huge elliptical galaxy M87 is slightly above the center. The galaxy cluster is estimated to contain about 1,500 galaxies.

A galaxy under influence

NGC 4911 is a remarkable spiral galaxy located within the Coma Cluster. There are conspicuous arms of gas and dust near the cluster, and much further away from the center, the galaxy is still surrounded by weak spirals of gas and stars. The thin structures are formed by gravitational disturbances from other galaxies in the densely populated galaxy cluster.

A cluster in the hair

At 320 million light-years, the Coma Cluster is much further from Earth than the Virgo Cluster; it comprises about 1,000 galaxies. The galaxy cluster is named after the rather modest constellation Coma Berenices (Latin for Berenice's hair). It's dominated by two huge elliptical galaxies, NGC 4874 and NGC 4889.

When you see a concentration of weak galaxies in the sky, it's hard to know if they really belong together. Perhaps the galaxies are at different distances, and we only see them randomly in more or less the same direction. Abell tried to take this into account by also observing the apparent brightness of the galaxies. Nevertheless, some of his galaxy clusters seem to actually be random constellations of smaller groups that have no relation to each other.

Elizabeth Scott and Jerzy Neyman tried to solve this problem mathematically. They subjected earlier galaxy counts to a thorough statistical analysis, which they published in 1958 – the same year that Abell's first galaxy cluster catalog appeared. Scott and Neyman proved that there aren't just numerous clusters of galaxies in the universe – there was no doubt about that – but also gigantic clusters of supergalaxies. Abell already had this idea himself (he called them "second-order clusters"), but he couldn't substantiate his theory.

Today we know that the Virgo Cluster is the central part of the gigantic Virgo Supercluster (also called the Local Supercluster), which is many millions of light-years across. The Local Group, to which our own Milky Way and the Andromeda Galaxy belong, is a relatively small concentration of galaxies in the outermost regions of this supercluster. Other large superclusters include the Centaurus Supercluster and the Perseus-Pisces Supercluster. In this way, there is a cosmic hierarchy: galaxies form small groups that in turn are arranged in clusters of galaxies, which in turn are part of gigantic superclusters.

In order to get an overview of the spatial distribution of galaxies in the universe, we don't just need to know the position of each galaxy in the sky but also their distance from us. Since the universe has been expanding since its birth (13.8 billion years ago), this isn't too difficult. The light waves of a distant galaxy are stretched on their long journey to Earth because the space in which they move is constantly expanding. The light arrives to Earth with a slightly longer wavelength than it had when emitted (and, therefore, with a color shift toward red). The greater this redshift is, the longer the light has been

A group photo

This cluster is located in the summer constellation Hercules and has several hundred galaxies. It sits about 500 million light-years from Earth. The Hercules Cluster contains many different types of galaxies, which also interact with each other. The image was taken with the European Southern Observatory's VLT Survey Telescope in Chile.

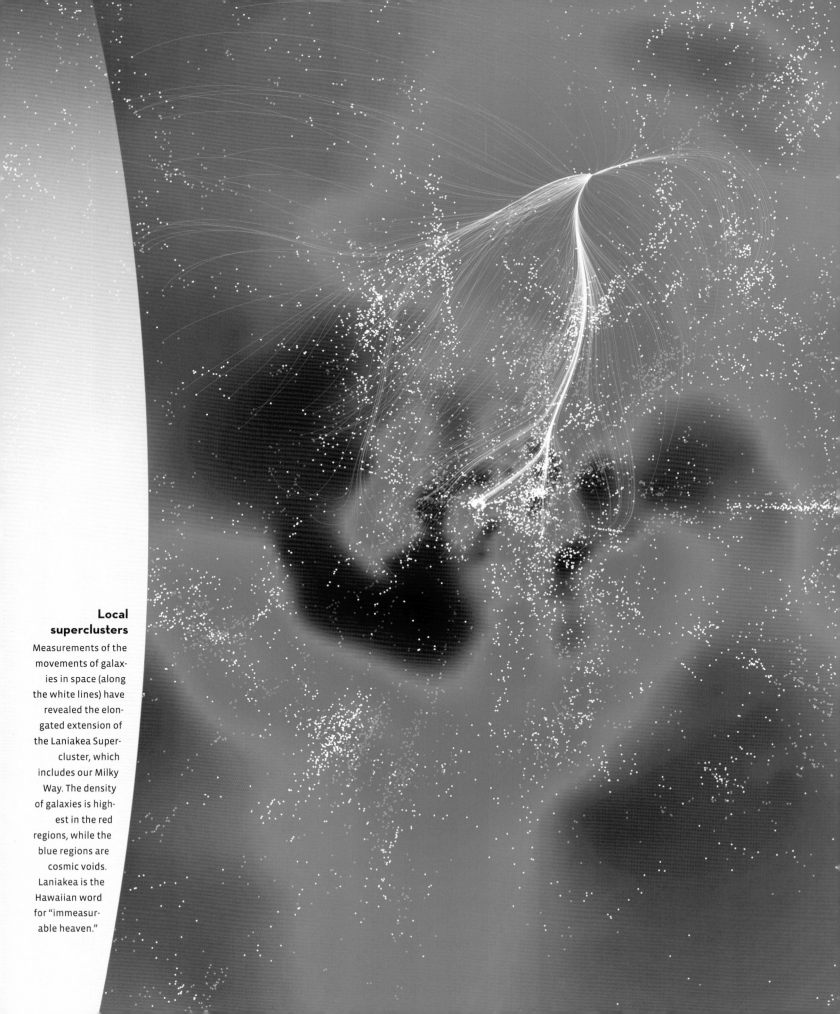

Local superclusters
Measurements of the movements of galaxies in space (along the white lines) have revealed the elongated extension of the Laniakea Supercluster, which includes our Milky Way. The density of galaxies is highest in the red regions, while the blue regions are cosmic voids. Laniakea is the Hawaiian word for "immeasurable heaven."

X-ray image
Like other galaxy clusters, the Perseus Cluster is filled with thin, hot gas that emits X-rays (shown in blue). The gas is concentrated around the central galaxy, NGC 1275. The radio radiation of this galaxy is reproduced in red tones. The intracluster gas is much heavier than all of the galaxies in the cluster combined.

traveling and the further away the source is. Maarten Schmidt also used this method in 1962 to determine how far away the mysterious radio star 3C 273 is from Earth.

Thanks to automated observation programs that simultaneously measure the redshift of dozens of galaxies, astronomers now have detailed 3-D maps of the distribution of galaxies in the universe. In addition, it's also been possible to measure the relative motions of galaxies, which allows researchers to record "flow patterns" that are the result of the gravitational effect of all the galaxy clusters and superclusters. These measurements helped confirm that the Virgo Supercluster is actually part of a much larger structure called the Laniakea Supercluster, named after the Hawaiian word for "immeasurable heaven" by its discoverers. Laniakea extends over half a billion light-years and contains at least 100,000 galaxies divided into several hundred individual groups and clusters.

If galaxies are the villages and cities of the universe, then clusters and superclusters are the urban metropolises, such as New York or Los Angeles. If there are large concentrations of galaxies, there must also be areas where they're much less numerous – the so-called voids (cavities that could be described as cosmic deserts). In 1981, American astronomers discovered the Boötes void, which is a gigantic empty area with a diameter of about 250 million light-years in which there are almost no galaxies. Since then, a large number of these cosmic voids have been discovered.

For a galaxy, it's significant whether it exists as a loner, like an oasis in a cosmic desert, or whether it's part of a large galaxy cluster. The odd galaxies that can be found here and there in the "empty" regions like the Boötes void lead quiet lives that remain undisturbed by environmental influences. But in a densely populated galaxy cluster, a galaxy is constantly exposed to these influences. The possibility of narrow transits or frontal collisions, for example, is much greater than outside a galaxy cluster – which is also the reason why there are so many elliptical galaxies in the central parts of galaxy clusters.

Moreover, the space between the galaxies within a cluster isn't really empty. The first X-ray telescopes launched into orbit around Earth revealed that clusters of galaxies are filled with extremely thin but also extremely hot gas. Because of its extremely high temperature – several tens of millions of degrees – this gas emits high-energy X-rays. A galaxy that moves at high speed through this "intracluster medium" can be blown almost completely clean and lose its gas supply, which limits the formation of new stars.

Gravitational Lenses

I grew up in a house from the 1920s, with jammed doors, leaking dormers, creaking roof beams and "war glass" windows. This inexpensive, rather bubbly glass certainly doesn't meet today's standards. For a curious and attentive little boy, however, it was fantastic: the bubbles in the glass worked like little lenses. When I sat at the dinner table, I often had one eye closed and moved my head back and forth so that the lamppost in the distance was directly behind such a glass bubble. And then I looked: the lens effect of the glass divided the lamppost in two. Magic.

The effect of a lens is always based on the curvature of light rays. Light moves slightly slower in water or glass than in air (or in a vacuum). This causes the light beam to change direction slightly as it passes the boundary between air and water or between air and glass at an angle. This effect is also familiar to most children: it's why a drinking straw appears bent in a glass of lemonade. With a round, polished lens, parallel light rays are refracted in such a way that they converge at one point – the focal point. This is the optical principle on which the effect of a classic telescope is based.

However, there are other methods of bending light. About 100 years ago, Albert Einstein calculated that light rays can also be bent slightly by the gravitational force of massive objects. Strictly speaking, light continues to follow a "right track" through four-dimensional space-time. However, this intangible substance is curved by the existence of matter, resulting in a light ray experiencing a slight deflection. The stronger the gravitational field and the smaller the distance in which a light beam passes through the object, the greater the deflection will be.

Einstein's prediction was spectacularly confirmed in May 1919, when astronomers made precision measurements of the positions of stars during a total solar eclipse. During such an eclipse, the bright surface of the Sun is completely covered by the Moon. Using a telescope, you can see stars near the Sun's disk that would otherwise be obscured by its light. Einstein theorized that if the light from these stars is actually bent by the gravitational force of the Sun, it should be possible to measure this using their observed position in the sky. The measurements from 1919 corresponded precisely with the predictions of general relativity. Einstein became world famous overnight.

It was only much later, in 1936, that Einstein published his ideas about gravitational lenses. Just imagine it with two stars. When viewed from Earth, one falls perfectly behind the other and is about twice as far away as the other. One would expect that the star in the background wouldn't be visible, but the light of this distant star, which hits the star in the foreground precisely and would therefore normally not reach Earth, is slightly bent by the gravity of the "lens star." As a result, it still arrives to Earth. If the Earth were exactly in line with these two stars, a ring of light would have to be visible around the foreground star – called an Einstein ring.

Einstein was also clear that such light rings could never be observed in reality. The chance that two stars are so precisely aligned with each other is enormously small. Since a star doesn't bend very much, the deflection effect is very minor and the Einstein ring remains very small. Other astronomers calculated that the same phenomenon must be observable for entire galaxies. These have much more mass and therefore cause a much stronger curvature of space. What's more, if the two galaxies aren't lined

Blackout practices

The deflection of starlight was first detected during a total solar eclipse in 1919, when the bright surface of the Sun disappeared behind the Moon and stars were visible in the background near the Sun. The silver-white corona is the Sun's thin gas envelope. This photo was taken from Spitsbergen during the solar eclipse on March 20, 2015.

Double vision
The two bright "stars" in the middle of this Hubble photo are two images of the same quasar – the nucleus of a distant galaxy. The quasar light reaches Earth along two different paths, so the distant object can be seen twice. A weak "lens system" is visible around the two images, which is responsible for the deflection of the light.

up exactly, a lens effect is still produced, even if the resulting image is less symmetrical.

In 1979, the first cosmic gravitational lens was discovered in the constellation Ursa Major. At the position where a powerful source of radio radiation had been found, there seemed to be not only one quasar (the bright nucleus of an active galaxy), but two quasars close to each other. Soon it became apparent that these were two images of the same distant object. On longer exposed photos, the (much weaker) lenticular galaxy was also seen. The gravitational force of the foreground galaxy split the quasar image in two, just as an imperfection in the window glass of my parents' house divided the distant lamppost in two. In the meantime, hundreds of comparable gravitational lenses have been discovered. Often you see two images (mostly from a distant quasar), but sometimes there are four. Complete Einstein rings have even been found, initially only in radio wavelengths but later also in visible light. For astronomers, gravitational lenses are veritable gifts from heaven: the image of a distant galaxy isn't just divided and distorted but also amplified, as is the case with a normal optical lens. Thanks to the gravitational lens effect of foreground galaxies, astronomers are able to study distant objects more precisely and in more detail.

The connection between gravitational lenses and galaxy clusters didn't become apparent until the 1980s. Independently of each other, French and American astronomers discovered unusual light arcs

Einstein ring

In the foreground of this photo there is a distant galaxy, about 10 billion light-years away, as seen from Earth just behind a massive galaxy. The gravity of the reddish foreground galaxy deforms the light of the distant blue object into an almost perfect ring. The existence of such gravitational lenses was first predicted by Albert Einstein in 1936.

A look into the depths

Abell 2218 is a beautiful galaxy cluster about 2 billion light-years from Earth, in the constellation Draco. This overview photo was taken with the 11¾-foot (3.6 m) Canada-France-Hawaii Telescope on Mauna Kea, Hawaii. The image also shows some foreground galaxies and a few stars in our own galaxy (at right).

in three distant galaxy clusters. Initially, no one had any idea which foreign objects were involved; the light arcs were most reminiscent of elongated "pearl necklace" galaxies. It turned out, however, that there was also a gravitational lens effect, in which the small image of a distant galaxy is pulled enormously lengthwise and intensified, not because of the gravity of a single foreground galaxy but because of the mutual gravitational field of the entire galaxy cluster.

The first spectacular image of the gravitational influence of a galaxy cluster was taken in 1995 by the Hubble Space Telescope. Hubble photographed the galaxy cluster Abell 2218, around 2 billion light-years from Earth, in the constellation Draco. Between the individual galaxies in the cluster are countless long streaks and concentric light arcs – the distorted images of distant background galaxies. Abell 2218 was later rerecorded using more sensitive cameras, and now many dozens of magnificent examples of galaxy clusters are known – depicting a distorted image of the distant universe as though we were seeing the world around us through the bottom of a jam jar.

Between 2013 and 2017, American astronomers carried out a time-consuming observation program with the Hubble Space Telescope, in which several carefully selected galaxy clusters were observed several times. The images obtained are the best – and most spectacular – examples of the gravitational lens effect of galaxy clusters. Among other things, this endeavor, called the Frontier Fields program, has led to the discovery of the farthest known galaxies – objects on the edge of the visible universe that would never have been detectable without the lens effect of the foreground galaxy cluster. In the last part of this book, I'll go into further detail about the investigation

Distorted images

The sharp eye of the Hubble Space Telescope shows countless light arcs in the center of Abell 2218. These are distorted images of galaxies in the background that are amplified and distorted by the gravity of the galaxy cluster. Abell 2218 was the first galaxy cluster in which this gravitational lens effect was so spectacularly photographed.

A cosmic lens

The distant and particularly massive MACS J0717.5+3745 galaxy cluster in the constellation Hercules was one of the objects of study in the Frontier Fields program conducted with the Hubble Space Telescope. The complicated gravitational field of the galaxy cluster provides distorted images of galaxies many billions of light-years from Earth.

of the very farthest (and very first) galaxies in the universe.

A surprising side effect of the Frontier Fields program was the discovery of a supernova in one of the "lensed" galaxies at a distance of about 9 billion light-years from Earth. Due to the complicated gravitational field of the foreground galaxy cluster (catalog designation MACS J1149+2223), this spiral galaxy has been imaged several times, and the light of the supernova reached Earth along four different routes at the end of 2014. At another point in the cluster we come across another image of the distant galaxy. This light had traveled a longer distance, and here the astronomers saw the same supernova flare up one year later, exactly as predicted.

The same applies to the double quasar mentioned earlier, where brightness fluctuations in one image only become apparent after some time (417 days to be exact) in the other image. This time difference is also caused by the expansion velocity of the universe, so measurements on gravitational lenses are therefore an independent (though not very accurate) method of determining this expansion velocity. Thus, research on galaxy clusters not only provides us with insights into the process of how the life cycle of a single galaxy is influenced by its environment, it also provides information about the properties and evolution of the universe as a whole – even about the mysterious dark matter in the cosmos, as we'll see in the next chapter.

Pandora's box

Abell 2744, in the constellation Sculptor, is also known as Pandora's Cluster. It was likely formed as a result of collisions between no less than four smaller galaxy clusters. The lens effect of the galaxy cluster makes it possible to explore the light of distant background galaxies that would normally be invisible or just barely visible.

Dark Forces

Galaxy clusters form the largest contiguous structures in the universe. A medium-sized cluster of galaxies can easily include hundreds or even thousands of galaxies, from large elliptical giants (often located in the center of the cluster) to smaller spiral galaxies (mostly found at the edges) and inconspicuous small dwarf galaxies. Furthermore, the space between the galaxies in a galaxy cluster isn't really empty: many intergalactic stars, planetary nebulae and star clusters have been discovered. In addition, galaxy clusters contain enormous amounts of extremely hot gas. Although it's very thin, this intracluster gas has more mass than all galaxies combined. Addtionally, Swiss-American astronomer Fritz Zwicky discovered during the 1930s that galaxy clusters also contain a great deal of invisible dark matter. Zwicky was a versatile astronomer who, among other things, invented the term "supernova" and – together with his colleague Walter Baade

Dusty galaxy
In addition to stars, interstellar gas and dark dust, galaxies also contain large amounts of dark matter, whose true nature is a mystery. This Hubble photo shows conspicuous dust clouds in NGC 1316, the central galaxy in the Fornax Cluster. However, dark matter within the universe doesn't consist of normal atoms and is therefore invisible.

A gravitational map
By measuring the distorted images of distant background galaxies, astronomers have been able to record the gravitational field of the foreground galaxy cluster MCS J0416.1-2403 (shown in blue). The observed weak lens effect can only be explained by assuming that the galaxy cluster contains large amounts of dark matter.

A cosmic shot put

The Bullet Cluster consists of two galaxy clusters that collided. The hot intracluster gas (shown in pink) has piled up between the two galaxy clusters, but the dark matter (in blue) is still distributed according to the individual galaxies in the two clusters. This distribution can't be properly explained by alternative theories of gravity.

– predicted the existence of neutron stars. In 1933, he researched the velocities of galaxies in the Coma Cluster. These galaxies moved so quickly that they would have long since flown out of the cluster if there hadn't been a strong gravitational field – much stronger than one would expect from the visible matter.

A year earlier in Leiden, Jan Oort had concluded that there must be invisible matter in the disk of our own Milky Way. Then Zwicky presented convincing evidence of the presence of gigantic amounts of dark matter in galaxy clusters. As explained earlier in this book, the existence of dark matter in galaxies was later convincingly proven by Vera Rubin and Kent Ford and by radio observations of rapidly rotating gas clouds. Measurements of the velocities of galaxies within galaxy clusters have also shown that such clusters contain a considerable amount of dark matter, even when accounting for the presence of hot intracluster gas.

Today, the presence of dark matter isn't just deduced from such speed measurements. In the previous chapter, we saw that the gravity of a galaxy cluster intensifies and bends the light of distant galaxies. It's also possible to calculate how much matter the galaxy clusters contain based on the gravitational lens effect. Every time, it appears that there is much more than just the visible matter perceived by optical telescopes (stars) and X-ray telescopes (hot gas). By closely observing the gravitational effect on the light of background galaxies, astronomers can even map the distribution of this dark matter. The elongated light arcs in some galaxy clusters are very noticeable. They're formed when the image of a distant galaxy is extremely distorted and elongated for whatever reason. In reality, however, every background galaxy is to some extent subject to the effects of gravitational lensing. In the early 1980s, American astronomer Anthony Tyson discovered that in some galaxy clusters, a conspicuous orientation of background galaxies can be observed, as though they were all stretched a little in the same direction (usually more or less concentrically around the center of the cluster). This effect is called weak gravitational lensing.

Of course, one never knows for sure whether a single elongated galaxy is caused by weak gravitational lensing or whether it actually has an elongated shape, such as a cigar-like elliptical galaxy or a spiral galaxy being viewed obliquely from the side. However, if you measure out the shape of dozens or hundreds of small background galaxies, you can

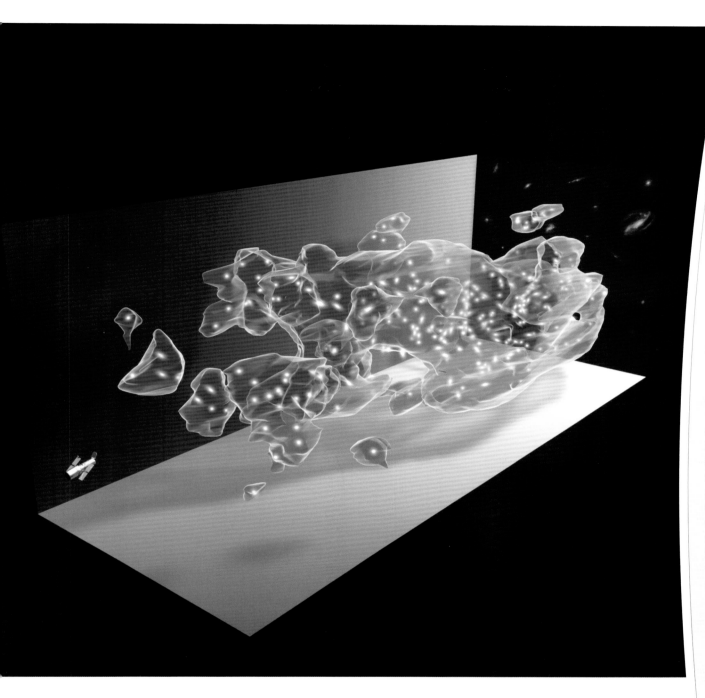

3-D image

By studying the effects of the weak lens effect at different distances in the universe, it has become possible to determine the three-dimensional distribution of dark matter. This illustration is based on data from the Hubble Space Telescope (left); the white spots are galaxies, and the blue clouds mark the distribution of dark matter.

determine if there is an unexpected preference for this direction. In this case, we can be almost certain that weak gravitational lensing is playing a role here and record the distribution of dark matter in the foreground galaxy cluster; it is simply a question of statistics.

However, not everyone is convinced of the existence of dark matter, since the conclusions are always based on the common notions of gravity. According to Newton and Einstein, the gravity of a celestial body decreases quadratically with its distance. At three times as great a distance from Earth, the gravitational force is only one-ninth. The theory of a modified Newtonian dynamics, or MOND (short for Modified Newtonian Dynamics), assumes that this ratio is not as accurate for weaker gravitational fields. In this case, we would draw false conclusions about the amount of matter present from the gravity measurements in the universe.

Observations made in 2004 on the now famous Bullet Cluster in the southern constellation Carina (Latin for keel of a ship), however, seemed to invalidate this alternative theory of gravity. The Bullet Cluster (officially called 1E 0657-558) consists of two

clusters of galaxies that have been colliding for hundreds of millions of years and even passed through each other. The galaxies themselves didn't experience a particularly strong impact (the possibility of collisions was minimal), but the hot gas within both clusters did collide. X-ray measurements show that, as expected, it piled up between the two galaxy clusters. The gas within a galaxy cluster always contains more mass than the galaxies within it.

However, measurements taken on the weak gravitational lensing of the Bullet Cluster show that most of the gravity is concentrated within or around the two individual galaxy clusters. This can't be explained with the MOND theory, which states that normal matter and gravity are equally distributed. With dark matter, however, the weak gravitational lensing of the Bullet Cluster can be explained quite well.

Dark matter consists of unidentified elementary particles that (apart from their gravity) hardly interact with each other. When the galaxies within the two colliding clusters are wrapped in extended halos of dark matter, the dark matter clouds also pass through each other. We expect that dark matter (and thus gravity) is distributed in the same way as the galaxies in the clusters.

Nevertheless, the mystery of dark matter in galaxies and galaxy clusters is far from solved. All observations of the mysterious substance are indirect. Experiments conducted in particle accelerators on Earth haven't yet been able to produce dark matter particles. Even measurements using sensitive underground detectors failed to detect the mysterious dark matter. Physicists have no idea what kind of particles they are; the only thing they know for sure is that the dark matter in the universe can't consist of ordinary atoms and molecules. For the time being, scientists seem to depend completely on astronomical observations to unravel this mystery. However, as I've said, they are not always unambiguous. Galaxy clusters aren't the only objects that exhibit weak gravitational lensing; this effect can also be observed in the vicinity of individual large galaxies. It's therefore also possible to record the halo of dark matter around such a galaxy.

In 1967, American astronomer James Gunn pointed out that the image of any distant background galaxy is more or less distorted by a weak lens effect, even if the light from that galaxy doesn't pass through a cluster of galaxies or narrowly past a foreground galaxy. This phenomenon is called cosmic shearing. By studying the statistics relevant to the forms of many thousands of distant galaxies, it's even possible

A mysterious ring

Measurements of the weak lens effect of a galaxy cluster (in this case, ZwCl 0024+1652, in the constellation Pisces) show astronomers the distribution of gravity in the galaxy cluster and thus the density of dark matter, shown here in blue. The origin of the conspicuous ring of dark matter at a great distance from the center of the galaxy cluster is not known.

A lightweight diffuser

The Hubble Space Telescope photographed this very faint galaxy (NGC 1052-DF2), which is 65 million light-years from Earth. It's as large as our Milky Way yet contains only 1/200 of the stars. Speed measurements on globular clusters also show that there is virtually no dark matter in this galaxy.

to create three-dimensional maps of the spatial distribution of dark matter in the universe. This is a way of practising cosmology that Fritz Zwicky never dreamed of. At that time, many hours of long exposure time were needed to photograph distant weak galaxies, and their properties had to be measured one at a time by hand. Today, large telescopes with extremely sensitive digital cameras record thousands of distant galaxies in just a few seconds and use intelligent computer algorithms to determine their position, dimension, shape and orientation. With the European Southern Observatory's VLT Survey Telescope's large OmegaCam camera and the National Science Foundation's Blanco telescope's Dark Energy Camera (both in Chile), the first major observation programs of cosmic shear have now been completed. In the future, much more light-sensitive measurements will be made by the Chile-based Large Synoptic Survey Telescope (currently under construction) and the European Space Agency's space telescope Euclid, which will capture the starry sky with the same sensitivity as the Hubble Space Telescope. So the 100-year-old mystery of dark matter may soon be solved.

A colorful witness

This photomontage of the galaxy cluster Abell 520, about 2 billion light-years from Earth, combines measurements in optical light (orange), X-rays (green) and dark matter (blue). In contrast to the Bullet Cluster, the dark matter in this cluster seems to be distributed in the same way as the cluster gas. This phenomenon can't be sufficiently explained with today's theories about dark matter.

The Large-Scale Structure of the Universe

The cosmos actually has an amazingly simple hierarchical structure, which might sound familiar to economists and business managers. A large international corporation consists of different companies, each with its own divisions. Each division is then divided again into different departments. Individual employees are only found at the "lowest" level. The same structure applies to the universe. Earth is one of the eight planets in orbit around the Sun, and the Sun is one of about 400 billion stars in the Milky Way. Galaxies are part of small groups and larger galaxy clusters; at the top of the hierarchy we find superclusters like Laniakea.

Superclusters were first discovered in the late 1950s, and the first large supervoid (the Boötes void) was discovered in 1981 – almost 40 years ago. However, it took some time for astronomers to systematically explore the large-scale structure of the universe. In fact, it didn't happen until the mid-1980s, when Margaret Geller, John Huchra and Valerie de Lapparent determined the positions and redshifts of several thousand galaxies in a narrow strip of the northern starry sky. As previously explained, the redshift of a distant galaxy is a measure of the time it took the light from that galaxy to reach Earth, so it is thus a measure of distance. Geller, Huchra and de Lapparent discovered that the measured galaxies aren't evenly distributed throughout the universe, and that there are long stretches of structures and relatively empty regions. One of these structures on the resulting map

A hint from God

The first 3-D map of the universe was produced in the mid-1980s. Our Milky Way is at the bottom, and the distances between galaxies (yellow dots) were determined by redshift measurements. The elongated structure is called the "Fingers of God"; it's created by the movements of galaxies in the Coma Cluster.

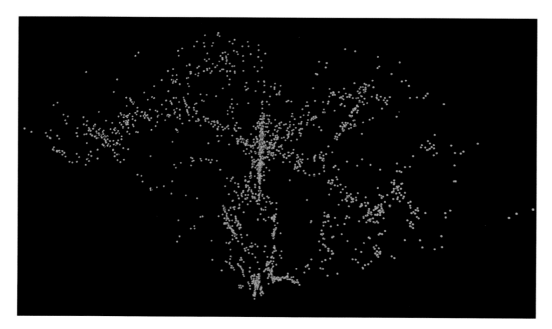

Computerized cosmos

This is an image from the EAGLE simulation, which recreates the evolution of the large-scale structure of the universe with a supercomputer. EAGLE stands for Evolution and Assembly of GaLaxies and their Environments. It's easy to see how matter first accumulates in fibrous striae and then flows to the nodes, where most galaxies are formed.

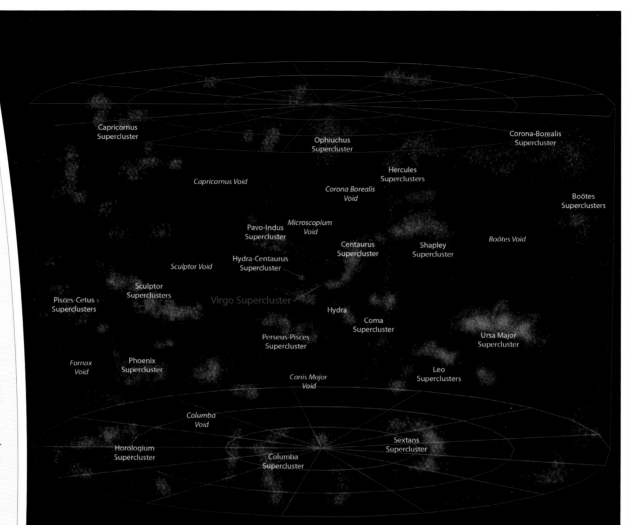

Spatial insight

This image represents the spatial distribution of galaxies and galaxy clusters. In it, the Milky Way is located exactly in the center, at the edge of the Virgo Supercluster. Its spatial position in relation to other superclusters (up to distances of half a billion light-years) is represented quite accurately.

A colorful map

In the image at right, the redshifts of – and distances between – hundreds of thousands of galaxies in the sky are shown in different colors, measured with the 2MASS Redshift Survey. The bright violet spot above is the Virgo Cluster; the light blue spot to the far left is the Perseus Cluster. The diagonal structure below the center of the map is the Pavo-Indus Supercluster.

appears to point like an outstretched index finger toward our Milky Way and was therefore given the nickname "Fingers of God." In fact, it is actually the well-known Coma Cluster. The elongated form is actually an illusion. It arises because the individual galaxies in the clusters have rather high velocities, which also affects the measured redshift – a "hint from God" that distance determinations in the universe aren't yet that simple.

However, what emerged quite clearly from these first measurements was the "Great Wall" – a gigantic, elongated collection of galaxies that measures about 500 million light-years long and about 200 million light-years high but is only 15 million light-years thick. Later, the Perseus-Cetus Supercluster complex was discovered elsewhere in the universe. It is also an extensive structure with dimensions of about one billion light-years. That discovery was followed in 2003 by the discovery of the Sloan Great Wall, which is 1.3 billion light-years long. In addition, astronomers discovered larger areas in the universe containing very few galaxies, such as the Giant Void (1.3 billion light-years in diameter) and the Eridanus Supervoid (1.8 billion light-years in diameter).

The first redshift surveys done by Geller, Huchra and de Lapparent were carried out with a 5-foot (1.5 m) telescope on Mount Hopkins in Arizona. It was a time-consuming task. Astronomers had to photograph the spectrum of each galaxy to determine the redshift. They later invented clever techniques to measure dozens of galaxies at the same time. In addition, digital detectors became increasingly sensitive, and astronomers were able to use larger telescopes. Over the decades, numerous redshift surveys have been completed, such as the 2dF Survey (with a large telescope of the Siding-Spring Observatory in Australia) and the Sloan Survey (with an advanced camera on a relatively small telescope in New Mexico).

Thanks to all these research projects, through which astronomers are penetrating ever further into the universe, cosmologists have now gained a good

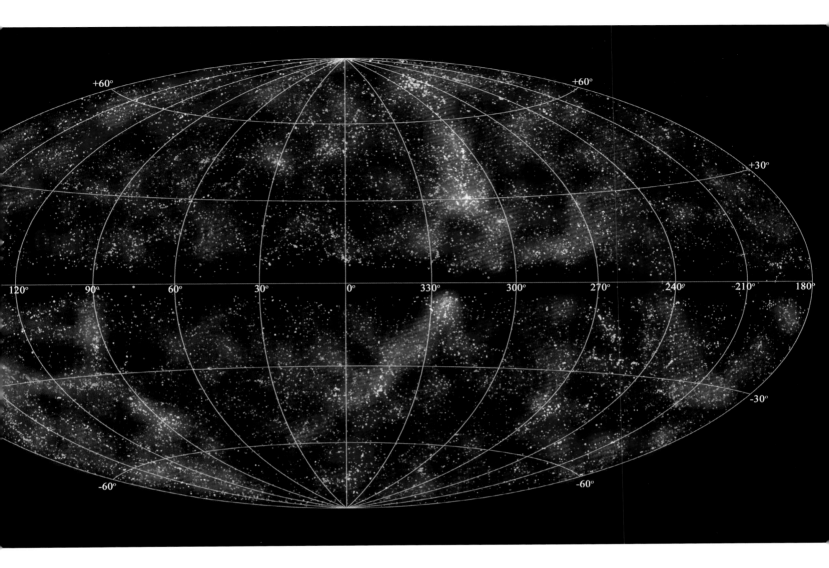

picture of the three-dimensional large-scale structure of the universe. This is comparable to the structure of soap foam: more or less spherical voids are surrounded by relatively thin walls in which the density of the galaxy is much higher. Where some of these walls meet, you'll encounter long filaments with even higher density, and at the intersections of these filaments you'll finds the largest and most densely populated clusters of galaxies.

Astronomers call this the "cosmic network" because the individual galaxies and groups are interwoven in fibrous structures with very thin and relatively cool gas. Even the cosmic network could be detected (albeit somewhat indirectly). The intergalactic gas leaves a kind of fingerprint in the light of distant quasars because the quasar light is absorbed in certain UV wavelengths.

One thing's for sure: the galaxies – the building blocks of the universe – are anything but evenly distributed. But how did this soap-foam-like large-scale structure come about at all? Shortly after the Big Bang, which occurred about 13.8 billion years ago, the universe was a rather homogeneous, hot "soup" of hydrogen and helium gas. The high-energy radiation emitted by the hot primordial soup can still be perceived as cosmic background radiation, which is actually the cooled residual heat of the Big Bang. Somehow the universe of today must have developed from a hot, uniform initial state, with fibers and filaments of galaxies.

At the end of the 1980s, it became apparent that the large-scale structure of the universe had been created through the effects of gravity. Small, random areas of denser gas could have attracted more and more matter over time and thus given the cosmos its lumpy structure. (These original density fluctuations are indicated by the minimal temperature differences of the background radiation.) However, 30 years ago it wasn't yet clear exactly how this process occurred. According to some cosmologists, the largest structures were formed first and then a kind of fragmentation took place in smaller clusters of galaxies and

Illustrious development

This image from the Illustris simulation shows the distribution of dark matter in blue tones and the distribution of normal matter (thin gas) in orange. These kinds of detailed computer simulations demonstrate that dark matter is the first to clump together to form a fibrous cosmic network; the regions with the highest density then form the most galaxies.

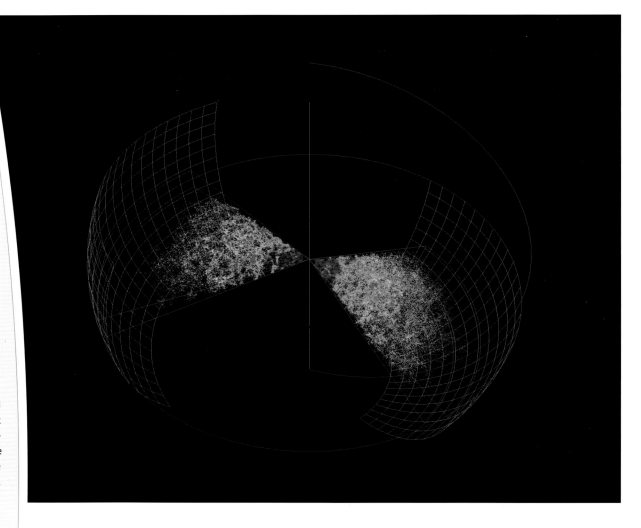

Cosmic pie slices

Thanks to the Australian 2dF Galaxy Redshift Survey, the spatial positions of tens of thousands of galaxies could be visualized in two flat segments up to distances of about 2.5 billion light-years. The density of galaxies is represented here in different colors; the long filaments and large superclusters can be clearly distinguished.

individual galaxies – the so-called top-down model. Other researchers thought the other way around, hypothesizing that individual galaxies had emerged first and then grouped into clusters of galaxies and superclusters at a later stage – the bottom-up model.

Today we know that this second theory best describes the reality. It can be seen, among other ways, from observations of the most distant and thus very first galaxies in the universe, which are described in more detail in the last part of this book. The bottom-up model is also supported by modern computer simulations that recreate the evolution of the universe in fast forward. This kind of computer simulation is based on a cube-shaped piece of cosmos that is filled with hydrogen and helium atoms, dark matter and radiation, which are the basic components of the universe. An important addition are small density fluctuations that are carefully coordinated with the measurements of cosmic background radiation. Afterward, this cube is stretched in all directions (the universe is expanding) and gravity starts taking effect.

The result is dark matter being concentrated in a fibrous pattern – the cosmic network – and normal matter flowing in the direction of the highest concentrations of density. The first stars and galaxies form quite quickly; at a later stage they produce thin walls, long fibers and densely populated galaxy clusters. After 13.8 billion years of accelerated cosmic history, the results of the computer simulations are amazingly similar to the real universe – an indication that we're on the right track, demonstrating the real origin and evolution of the large-scale structure.

Reproducing the evolution of the entire universe with a computer sounds quite ambitious. Moreover, you need an almost unimaginably powerful computer. The first modest attempts were thus quite primitive. But as powerful supercomputers became more affordable, the quality of the simulations improved steadily. In the latest calculations, even complicated hydrodynamic processes are taken into account in order to describe the behavior of the gas as accurately as possible, even on a small scale, and to adjust to the level of individual galaxies.

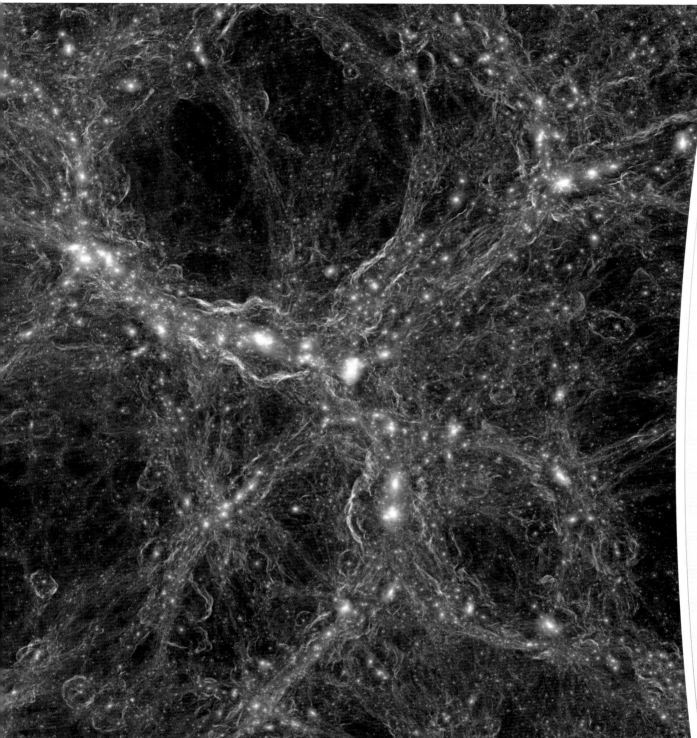

Shocking images
The latest computer simulations of the evolution of the universe demonstrate not only the process of dark matter clumping together (blue tones) and the birth of galaxies (the yellow regions) but also the shock waves generated in the thin intergalactic gas. This picture is taken from the Illustris TNG simulation (TNG stands for The Next Generation).

In recent years, research into the evolution of the universe and the formation of galaxies has increasingly become a symbiotic relationship between experiment and theory and between observations and simulations. The cosmic background radiation is in some ways a baby photo of the universe, while around us we see the universe at its current age. There also seems to be only one theoretical model that combines observations and simulations. This standard cosmological model, on which all these impressive computer simulations are based, account for not only dark matter but also large amounts of mysterious dark energy. This enigmatic part of the universe, however, is reserved for the last part of this book.

INTERMISSION

A Glance into the Past

When looking at the universe, we're not only looking far into space but also far back in time. NGC 7331 is a spiral galaxy in the constellation Pegasus. It's 45 million light-years from Earth. This means that it took 45 million years for the light from this galaxy to reach Earth. We thus see the galaxy as it looked 45 million years ago, during the Eocene epoch, when Australia separated from Antarctica and when Europe and North America began to drift apart. There were no traces of the earliest forerunners of humans.

By observing distant galaxies, astronomers are able to learn more about the history of the universe. The further you look, the further you travel back in time. Telescopes are actually time machines.

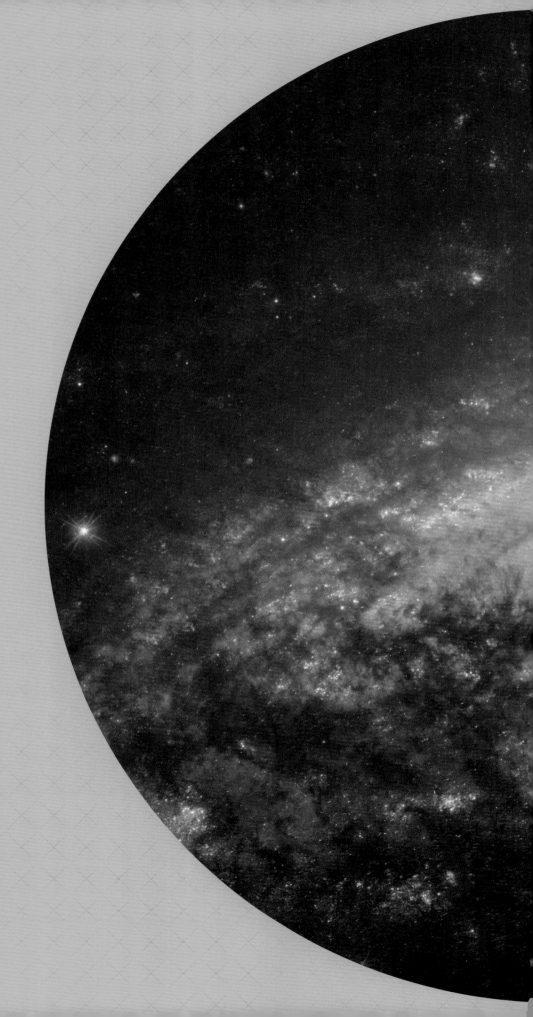

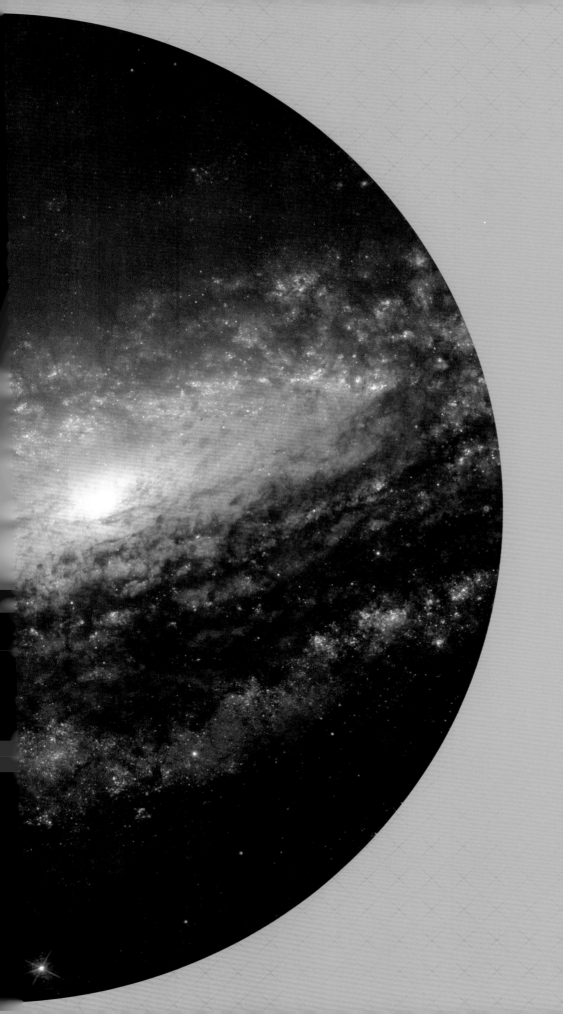

A heavenly sample

This photo of a small part of the starry sky in the constellation Scutum was taken by the Hubble Space Telescope as part of the Frontier Fields program. By photographing different, randomly selected fields, astronomers get a good picture of cosmic variance, which is the measure of deviations from the average.

Birth and Evolution

At the Edge of Space

Although no one has ever been able to count them accurately, the number of galaxies in the universe is estimated to be several hundred billion. That's a few dozen galaxies for every person on Earth. It's an unimaginably great number with an equally unimaginably great diversity, as we've already seen.

Unfortunately, it's not easy to observe all these galaxies in detail, particularly due to the enormous distances between them and Earth. To give you an idea, the Andromeda Galaxy is "only" 2.5 million light-years away from us, while most galaxies are several hundred or even several thousand times further away.

A look into the future

The 21¼-foot (6.5 m) James Webb Space Telescope, to be launched in 2021, is considered the successor to Hubble. The Webb telescope will also be used for long exposures of small parts of the starry sky, mainly in infrared wavelengths. This computer simulation gives us an idea of what astronomers expect from this Webb Deep Field.

Diving into the depths

The first Hubble Deep Field shows about 2,000 galaxies from a small region of the starry sky in the constellation Ursa Major. A total of 141 hours of exposure were used to make even the most distant and weakest small galaxies visible. The strange notch at the top right was caused by Hubble's Wide Field and Planetary Camera 2, which was used to take the pictures.

Lights, camera, action! The Hubble Space Telescope's Advanced Camera for Surveys, installed by shuttle astronauts in 2009, identified some 10,000 extremely distant galaxies in the Hubble Ultra Deep Field in the constellation Fornax (chemical furnace). This photomontage combines images in visible, ultraviolet and infrared lights.

Beyond the limit

In the center of the galaxy cluster Abell S1063, in the southern constellation Grus, the gravitational lensing effect is so strong that inconspicuous background galaxies become visible for the cameras of the Hubble Space Telescope. This photo was taken as part of the Frontier Fields program.

Under the magnifying glass

A distant cluster of galaxies dominated by two bright elliptical galaxies forms a cosmic lens for objects at much greater distances from Earth. This is how the elongated images of these background galaxies are created. The gravitational lens effect makes it possible to observe these distant objects in detail. The star on the left is a foreground star in the Milky Way.

Nevertheless, the research on these enormously distant galaxies is very important, and not just because it may lead to a new entry in the *Guinness Book of World Records*. The light of a galaxy 10 billion light-years away takes 10 billion years to reach Earth. We therefore see this galaxy as it looked 10 billion years ago, when the universe was still less than 30 percent of its present age. Looking far into space automatically means looking far back in time. Research on distant galaxies thus makes it possible to observe their cosmic evolution.

Since light has a limited speed, we can only perceive a limited part of the universe. The universe was created 13.8 billion years ago, which means that a beam of light can't have traveled for more than 13.8 billion years. At a distance corresponding to a light-travel time of 13.8 billion years, this "observation horizon" terminates a deeper view into space. Behind this horizon there are still countless galaxies whose light hasn't yet reached us. The several hundred million galaxies mentioned above thus refers to the observable universe. No one knows how far the cosmos extends beyond this horizon – perhaps infinitely far.

Shortly after the launch of the Hubble Space Telescope in April 1990, when the instrument was somewhat "short-sighted" due to a manufacturing defect in the main mirror, American astronomer Mark Dickinson took some very long-exposed photographs of small sections of the starry sky. In these Hubble photos, extremely weak spots of light were visible – galaxies 8 to 9 billion light-years away. For the most part, these were not perfectly symmetrical spiral galaxies but rather small, irregularly shaped dwarf galaxies. Hubble allowed astronomers to glimpse into the youth of the universe and the early evolution of galaxies. When the Hubble telescope was finally equipped with a corrective instrument, there was every reason to repeat Dickinson's photo.

The cosmic evolution

A relatively large part of the starry sky has been studied in detail by various telescopes, including Hubble, as part of the Great Observatories Origins Deep Survey (GOODS). The resulting image penetrates less deeply than the Hubble Ultra Deep Field, but it covers a larger area of the sky. This is how astronomers study the evolution of galaxies.

Under the direction of Bob Williams, then director of the Space Telescope Science Institute in Baltimore, a unique project was carried out that became known as the Hubble Deep Field. At the end of December 1995, Williams and his colleagues selected an apparently empty part of the starry sky in the constellation Ursa Major and took a total of 342 pictures of it with a total exposure time of 141 hours. Some astronomers were strongly against this project, believing it was a waste of precious Hubble time and would provide few valuable insights. However, the results exceeded all expectations.

Approximately 2,000 individual galaxies were visible in the original Hubble Deep Field photo! Some were relatively large and bright, such as the striking spiral galaxy in the lower left corner of the image. Some were elliptical galaxies, such as the one in the upper half of the image. Most, however, were tiny and misshapen. If you let your gaze wander slowly over the photo, you can imagine that you are making a journey of hundreds of millions of light-years through the wide universe. When you then also realize that all these small dots of light are complete galaxies, often with many billions of stars, then you can't help but be impressed at the size of the universe.

Large sensitive terrestrial telescopes, such as the 33-foot (10 m) Keck telescope on Mauna Kea in Hawaii, were used to study most of the galaxies in the Hubble Deep Field in detail. The Keck telescope was able to determine the spectrum of the brightest specimens and actually measure their redshift – a direct measure of that galaxy's distance from us. Astronomers could not take such measurements of the weakest galaxies, but by measuring the brightness of these galaxies in different colors, they could get quite reliable indications of the redshift. Thus, astronomers transformed the "flat" Hubble Deep Field photo into a three-dimensional map of a small part of the starry sky – a kind of cosmic borehole core.

After the success of the Hubble Deep Field project, the cosmologists had gotten the hang of it. In the autumn of 1998, a comparable photograph was taken of a small part of the starry sky in the southern constellation Fornax (chemical furnace) – the Hubble Deep Field South project. This was done not only in visible wavelengths but also in infrared light. Hubble Ultra Deep Field (2003/2004) and Hubble Extreme Deep Field (2012) followed later, once the space telescope was equipped with new, more sensitive cameras (which also had a slightly larger field of view). Other projects brought bigger parts of the sky into the picture, but with shorter total exposure times. Of course, the various Hubble Deep Fields have also been studied in detail by space telescopes such as Chandra and Spitzer, allowing observations to be made in X-ray wavelengths and in the far infrared.

These projects all had the same goal: to learn as much as possible about the evolution of galaxies, which is only possible by looking far back in time. All these observations have made it clear that the universe has undergone a profound evolution over its first billion years.

The first irregularly shaped galaxies emerged relatively quickly. Since the density of the universe billions of years ago was noticeably greater than it is today, many more collisions and mergers occurred at that time. For example, the first protogalaxies clumped together to form ever-larger specimens. About 11 billion years ago, the birth rate of new stars reached its peak; after that, the rate of star formation gradually slowed down. Since the American-manned space program was discontinued, maintenance flights to the Hubble Space Telescope can no longer be performed. It's thus no longer possible to replace today's cameras with even more sensitive devices. Astronomers, however, have come up with a trick that, in principle, allows them to look even further back in time. As part of the Frontier Fields program, which ran from 2013 to 2016, the space telescope was aimed at six distant clusters of galaxies known to exert a strong gravitational lensing effect on light from distant background galaxies. This "natural" telescopic effect allows Hubble to detect objects that would otherwise never be visible.

In addition to these six galaxy clusters, six randomly selected parts of the starry sky were observed in exactly the same way. This could be seen as a kind of control sample. Ultimately, the universe looks about the same in every direction, although it's never identical. The additional fields are necessary to get a good picture of cosmic variance.

The observations in the Frontier Fields haven't yet all been fully evaluated. This work is very time-consuming, as the strongest gravitational lens effects take place in the center of the galaxy cluster, where the disrupting light from foreground objects hinders visibility. What's clear, however, is that this really was the last thing out of Hubble. Only in 2021, after the launch of the James Webb Space Telescope, will astronomers be able to penetrate even further into the cosmic past.

This does not, however, mean that nothing at all is known about the birth of the very first galaxies in the universe. Thanks to observations in infrared and millimeter wavelengths, it's been possible to observe galaxies at such great distances that their light must have traveled to Earth for more than 13 billion years. We're thus looking back into a period when the universe was less than 800 million years old. Of course, not all of this is easy. But thanks to research on these very first galaxies, astronomy is almost at the cradle of the universe.

The First Galaxies

Old light

NGC 1015 is a large barred spiral galaxy in the constellation Cetus. The measured redshift shows that it's 120 million light-years away from us. The light we receive from this galaxy here on Earth was emitted 120 million years ago, during the Cretaceous period, when Earth was populated by dinosaurs.

It will always be difficult to imagine the extent of the cosmos. Man has never been further than the Moon – a journey of a few hundred thousand miles, equal to about 10 Earth orbits. Through unmanned space probes, we've explored other planets at close range, and some of these planetary probes are currently leaving the solar system. But this means very little on a cosmic level: 12 billion miles (20 billion km) (the distance that *Voyager 1* had covered up to early 2018) is only one-twentieth of a percent of the distance to the nearest neighboring star.

Since the speed of light is the fastest possible speed in nature (at nearly 190,000 miles per second (300,000 km/s)), astronomers often express distances in the time it takes a beam of light to travel that distance. This is less than 1.5 seconds to the Moon, a little over 8 minutes to the Sun and about 6 hours to the distant dwarf planet Pluto. Radio signals from the *Voyager 1* space probe, which are also traveling at the speed of light, take almost 20 hours to reach Earth; a beam of light takes more than four years to travel to the neighboring star Proxima Centauri.

But in the world of galaxies, even one light year is insignificant. Light from the Andromeda Galaxy is still traveling 2.5 million years to Earth at the unimaginably high speed of 5.9 trillion miles (9.5 trillion km) per year. Other galaxies are tens or hundreds of millions of light years away. When you read (or write) a lot about such tremendous distances, you finally start getting used to it. However, no one can truly get a real idea of such gigantic distances.

It's often helpful to realize that looking far into space invariably means looking far back into time. A radio signal from *Voyager 1* that reaches Earth today was transmitted yesterday from the distant space probe. Light that we received in 2018 from Proxima Centauri started its journey in 2014. If you look at the Andromeda Galaxy on a clear autumn night, you will see photons that began their journey to Earth 2.5 million years ago, when *Homo habilis* was first using primitive stone tools.

The distance from Earth to the Virgo Cluster is 65 million light-years. We see the galaxies in this cluster as they looked when the dinosaurs died out. And yet, 65 million years into the past still doesn't cover more than half a percent of the lifetime of the universe. This reveals the challenges astronomers face when they look back to the earliest years of the universe. These are galaxies that are so far away that their light has taken many billions of years to reach Earth.

Cosmic time spans are probably just as unimaginable as cosmic distances. The universe has existed for almost 14 billion years. This can be compared to an encyclopedia consisting of 14 thick volumes each with 1,000 pages. In such a framework, the Sun and Earth would have emerged halfway through volume 10, the dinosaurs would have died out on page 935 of volume 14, *Homo sapiens* would have popped up somewhere in the lower half of page 1,000 of volume 14, and the history of mankind would be written in the second half of the very last sentence. Astronomers are actually trying to look back into the first volume of the encyclopedia, when the very first stars and galaxies emerged.

Under favorable conditions, the Andromeda Galaxy is visible to the naked eye. However, the apparent brightness of a celestial body decreases quadratically in relation to its distance from the viewpoint: at two times the distance, the brightness is only one-fourth, and at three times the distance it's only one-ninth. If the Andromeda Galaxy were 10 billion

A baby galaxy
An extremely distant, very dusty galaxy was discovered by the ALMA observatory in Chile. We see the A2744_YD4 galaxy as it looked when the universe was barely 600 million years old. This illustration shows that the newborn galaxy is almost without structure and that new stars form in it very quickly.

A varied view
On this Hubble photo you can see diverse galaxies that are located at various distances from Earth. The large galaxies are relatively close together, but the photograph also shows innumerable inconspicuous small points of light – newborn galaxies from the early days of the universe, many billions of light-years from Earth. The bright orange star is a foreground star in the Milky Way.

light-years away instead of 2.5 million light-years (4,000 times as far), it would be 16 million times weaker. Extremely light-sensitive instruments are needed to observe the most distant galaxies in the universe, where astronomers also look far back in time.

As already explained, due to the expansion of the universe, the light waves of a distant galaxy are stretched on their way to Earth: they reach Earth with a longer wavelength than that with which they were emitted. Their light will have shifted toward red; this redshift – a reliable measure of distance – can be measured by decomposing the galaxy's light into spectral colors and examining the resulting spectrum in detail. However, if the galaxy is extremely weak from the beginning, no signal remains after the light is split. In cases such as this, distances cannot be measured using the light spectrum.

Instead, astronomers use what is sometimes called the dropout technique. A distant, faint galaxy is observed through different colors of filters, and the brightness of the galaxy is measured in these different wavelength ranges. Radiation with a wavelength of less than 91.2 nanometers (ultraviolet) is absorbed by neutral hydrogen gas in the galaxy itself and its immediate surroundings. Therefore, only radiation with a wavelength of more than 91.2 nanometers will reach Earth. But if all this radiation undergoes a high redshift, we will only see light on Earth with a longer wavelength than, for example, 600 nanometers (orange). When viewed through a red filter, the galaxy is still clearly visible, but it can no longer be seen in photos taken with a yellow, blue or ultraviolet filter. The wavelength at which this dropout – that is, when the galaxy disappears from the image – occurs is an approximate indication of its distance from Earth.

Incidentally, there is often another effect involved. When a galaxy contains large amounts of dust, the starlight is absorbed by the dust clouds. The dust heats up, radiates infrared radiation and makes the galaxy brighter in the infrared range.

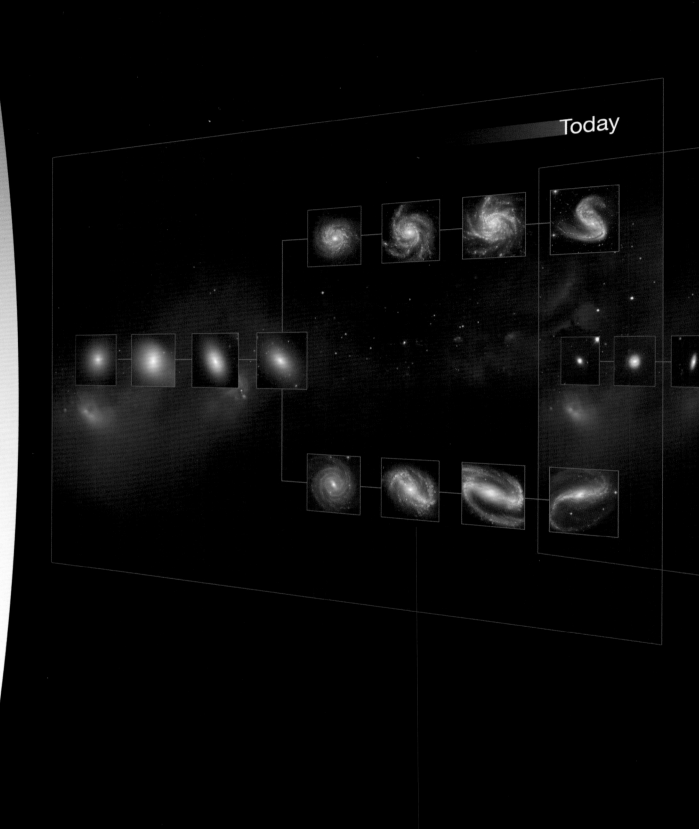

Cosmic evolution

In the famous tuning fork diagram by Edwin Hubble (left), the galaxies are arranged according to their shape and structure: elliptical galaxies (far left), spiral galaxies (above) and barred spiral galaxies (below). Astronomers have studied all of these different types of galaxies and determined what they probably looked like billions of years ago (center and right).

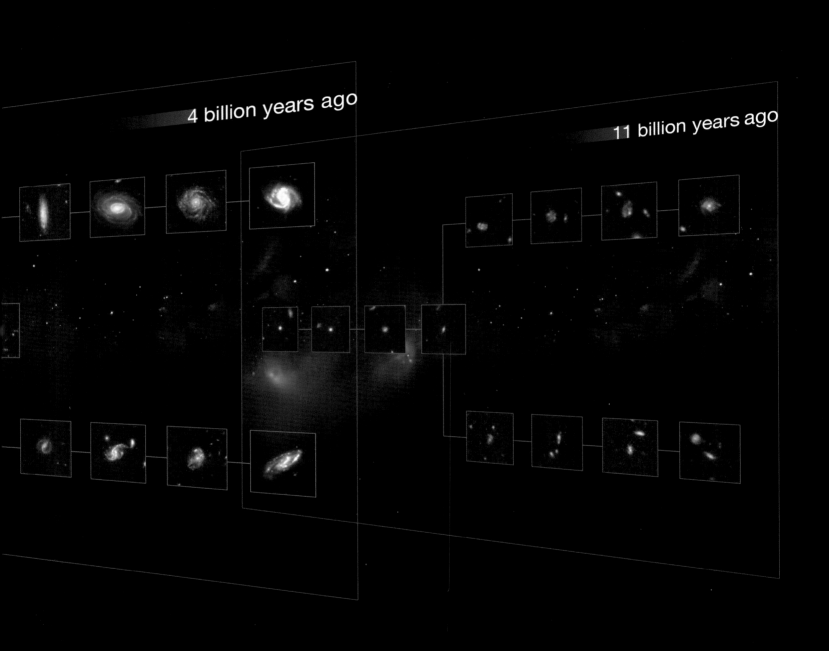

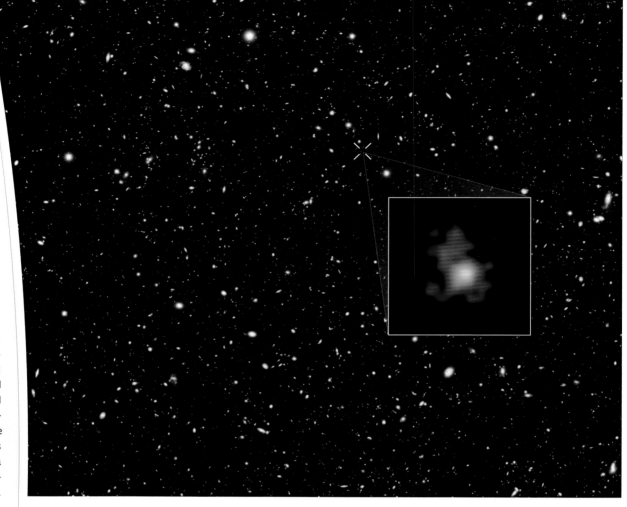

A distance record

As of spring 2016, GN-z11 is the most distant galaxy known, at a distance of 13.4 billion light-years from Earth. The formless clump of gas and stars is visible with the Hubble Space Telescope. Light from this protogalaxy started its journey toward Earth when the universe was no more than 400 million years old. The red color is a result of the enormous redshift.

If a galaxy is very far away, meaning that the emitted light has undergone a strong redshift, this infrared radiation arrives on Earth as microwave radiation with a wavelength in the order of 1/32 inch (1 mm). Such galaxies can't be seen at all with normal telescopes. Astronomers are using a system of large parabolic dishes, such as the Atacama Large Millimeter/submillimeter Array (ALMA) in northern Chile to locate this microwave radiation.

Unfortunately, this research doesn't provide us with beautiful photos. The distant galaxies look like inconspicuous blurry spots of light that don't really tell you anything. In many cases, these are small, irregularly shaped objects. Some are no more than a few hundred light-years in diameter and have less than one percent of the mass of the Milky Way. In many ways, they can be compared to large star-forming regions in today's galaxies, such as the Tarantula Nebula in the Great Magellanic Cloud. Similarly, when the universe was in its infancy, the birth of stars had just begun, and the cosmic "primordial soup" had just begun to densify, so very few galaxies had merged. Our galaxy was formed billions of years ago from precisely such formless gas and star clumps.

Despite the enormous progress in exploring the very first galaxies in recent years, there are still many unanswered questions. Astronomers have discovered quasars, the extremely bright nuclei of galaxies, at distances of many billions of light-years from Earth. It is generally believed that quasars derive their energy from the presence of a supermassive black hole, but it is not clear how such black holes formed so quickly during the evolution of the universe. Even the origin of large amounts of dust in some of these early galaxies isn't easy to explain: shortly after the Big Bang, the universe consisted mainly of hydrogen and helium.

We can hope that observations using the next generation of telescopes – the James Webb Space Telescope and the Extremely Large Telescope – will shed new light on the earliest phase of the universe and thus on the formation of galaxies.

Early whirlpools

Measurements taken by the ALMA observatory in northern Chile have shown that some galaxies had a beautiful, ordered structure shortly after their formation. This illustration shows a newly formed galaxy, almost 13 billion light-years from Earth, which has been found to rotate in the same direction as the Milky Way.

The Dawn of the Universe

In the beginning, God created Heaven and Earth. This is the first sentence of the biblical story of creation. Short, simple and clear. For centuries, everyone was content. Even in the 17th and 18th centuries, most astronomers were convinced that the question of the origin of the universe was found in the realm of religion, not science. As the domains of religion and science drifted further apart, people often assumed for the sake of simplicity that the universe had always existed – Albert Einstein, for example, still believed this a hundred years ago.

The discovery of the expanding universe at the end of the 1920s, however, put an energetic line through this cosmic permanence. If the distances between galaxies are increasing today, they must have been much smaller a long time ago. In 1931, Belgian astronomer and Jesuit priest Georges Lemaître was the first to present a scientific alternative for the history of creation: the cosmos was created from a kind of "uratome" or "cosmic egg" with an unimaginable density and temperature. Today, Lemaître is generally regarded as the spiritual father of the Big Bang theory.

First stars

The first stars in the universe produced large amounts of ultraviolet radiation. Through the energy of this radiation, the cold, neutral hydrogen and helium gas in the universe (shown in red) was gradually re-ionized (shown in blue), as demonstrated in this artistic impression. We still only know very little known about this particular period of re-ionization in cosmic history.

On the run

As a result of the expansion of the universe, the distance between our Milky Way and NGC 3621 increases by about 300 miles (500 km) per second. This galaxy, here in a photo taken with the European Southern Observatory's 7¾-foot (2.2 m) telescope in Chile, is currently 22 million light-years away from us. It will be hundreds of millions of years before the distance has increased to 23 million light-years.

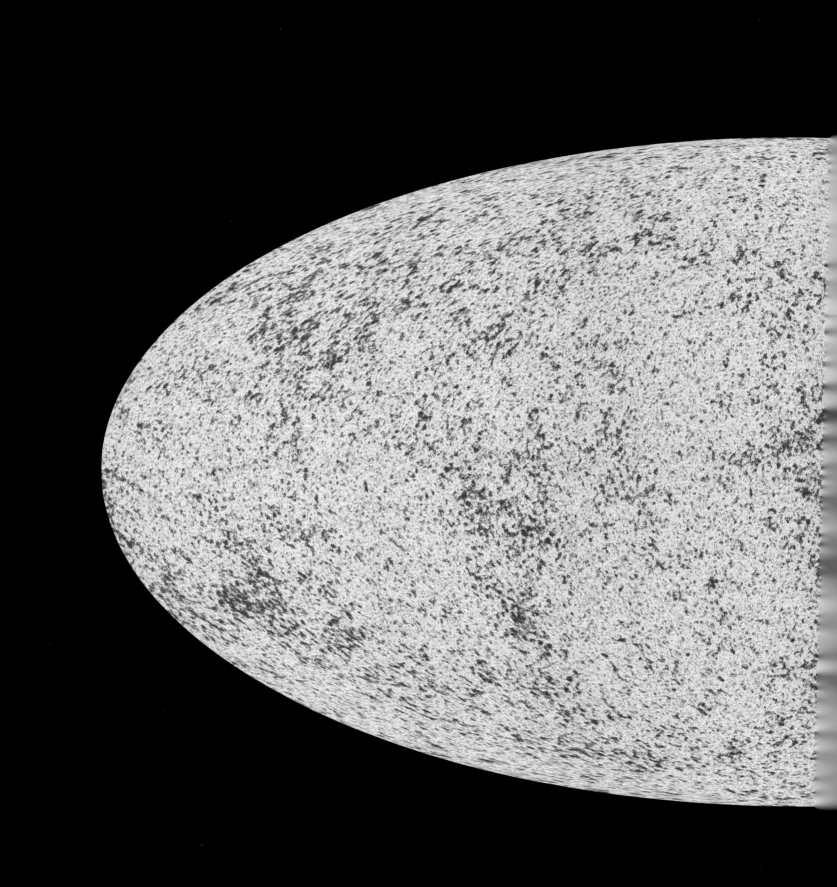

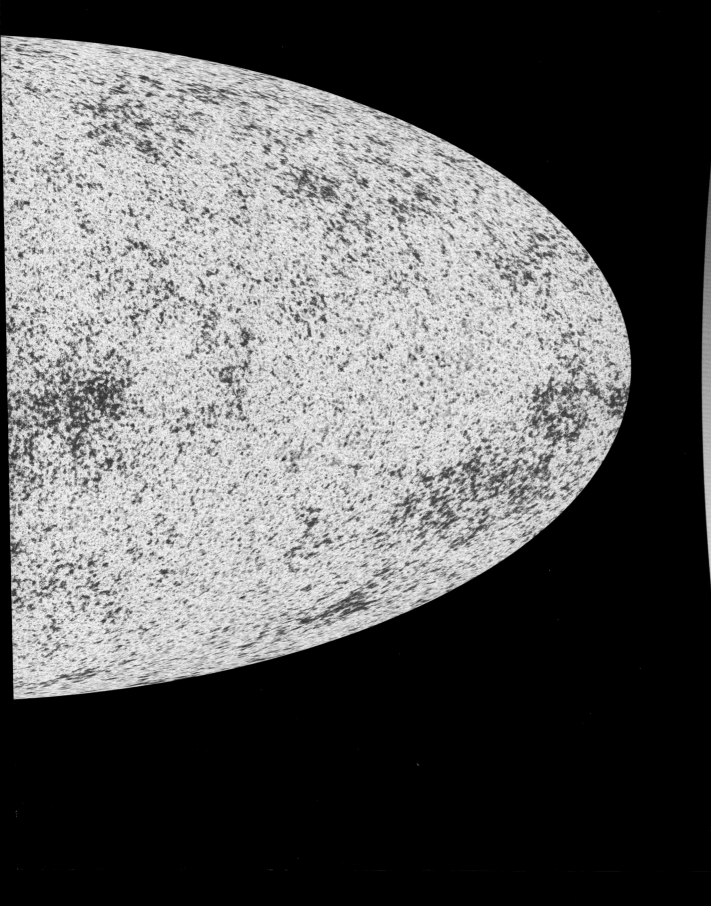

A baby photo

The red and blue spots on this map of the starry sky reflect tiny temperature differences in the cosmic microwave background radiation as recorded by the European Space Agency's Planck Space Telescope. These weak temperature differences are the result of small density fluctuations that took place in the newly formed universe. These areas later became galaxies.

Growth spasms
This is an artistic representation of two colliding galaxies during the early days of the universe, as observed by the ALMA Observatory in northern Chile. Such collisions and subsequent mergers occurred much more frequently in the distant past as compared to today; they triggered the formation of large galaxies such as the Milky Way.

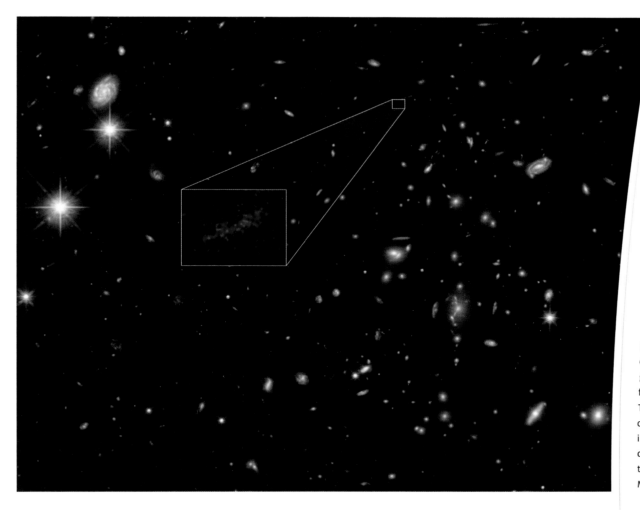

An embryonic galaxy

SPT0615-JD is one of the most distant galaxies ever observed. The strongly redshifted protogalaxy is visible thanks to the gravitational lens effect of a massive galaxy cluster in the foreground (at right). The early galaxy is only 2,500 light-years in size and weighs only one percent of the mass of the Milky Way.

There are many misunderstandings circulating about the Big Bang. Most people imagine the birth of the cosmos as a kind of explosion in empty space, but this concept is wrong. The Big Bang didn't happen at one point in space but everywhere at the same time. When the universe was created, the whole of space was a boiling sea of energy. The birth of the universe can therefore be better described as an explosion of space rather than an explosion in space. Moreover, this first beginning gives rise to many puzzles. Scientists still don't know whether the Big Bang itself had a cause and whether time also originated in the Big Bang.

The bubbling energy of the Big Bang gave rise to pairs of particles and antiparticles that immediately extinguished each other by emitting gamma rays – all according to Einstein's famous formula $E = mc^2$. Yet for some reason there was a small amount of matter left. The newborn universe consisted of high-energy photons (light particles) and a mix of elementary particles: protons, neutrons, electrons, neutrinos and mysterious dark matter particles. About three minutes after the Big Bang, atomic nuclei of helium (made up of two protons and two neutrons) were formed, and the universe roughly consisted of three-quarters hydrogen nuclei (protons) and one-quarter helium nuclei.

Since atomic nuclei have a positive electrical charge and electrons are negatively charged, we often refer to plasma, which is a mixture of electrically charged particles, in relation to them. Light can't move unhindered in plasma, which means that plasma is opaque, like the flame of a candle. After the first 380,000 years, the expanding universe had cooled sufficiently to form uncharged atoms: negatively charged electrons connected to positively charged atomic nuclei. After this connection, the primordial soup was electrically neutral, and the remaining radiation from the formation phase of the universe could spread unhindered into space.

Almost 14 billion years later, this cosmic background radiation is still present everywhere in the universe. Its intensity has, understandably, decreased enormously, and its wavelength has been greatly stretched as a result of the expanding universe. The intensity peak of the background radiation is no longer in the visible part of the light spectrum but rather in the microwave range, at a wavelength of

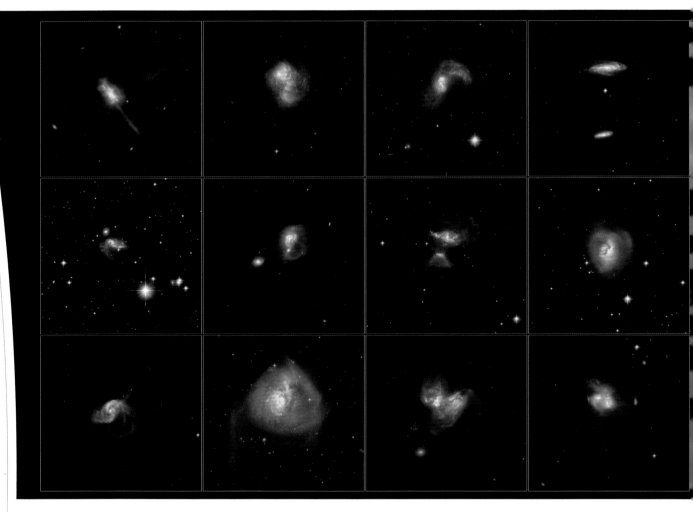

Victims

The Hubble Space Telescope has revealed dozens of spectacular examples of galaxies passing or even colliding with each other at close range. Such encounters, in which the two galaxies are deformed by mutual attraction, were much more frequent during the early days of the universe, when the universe hadn't yet expanded as much.

1/32 inch (1 mm). The corresponding radiation temperature is only 2.7 degrees above absolute zero. It's no wonder that cosmic background radiation wasn't discovered until 1965!

The temperature distribution of background radiation in the starry sky says something about the properties of the universe at the time the radiation was first released, about 380,000 years after the Big Bang. If all matter had been distributed evenly in space, the temperature of the background radiation would have shown exactly the same value everywhere. In that case, galaxies, stars and planets would never have formed in the universe. Instead, we see minimal temperature fluctuations in the background radiation, caused by subtle density differences in the young universe. Small areas, in which matter sat just a little closer together than average, became areas where galaxies later formed.

However, it would be more than 100 million years before the first stars and galaxies saw the light. To put it another way, before the cosmos was born again. The matter in the newborn universe was initially glowing hot and radiant, just as dazzlingly as the Sun. But the gas quickly cooled down to such an extent that it no longer emitted any visible light. The mysterious dark matter in the universe initially began to clump together more and more into a streak-like three-dimensional network. Hydrogen and helium atoms, influenced by gravity, flowed into the regions with the highest density of matter.

This whole process took place in the dark, during the dark age of the early universe. At some point somewhere in this expansive darkness the very first star must have ignited when a small concentration of hydrogen and helium gas contracted so strongly under its own gravity that nuclear fusion reactions began. Shortly after, the lights came on everywhere else in the universe.

Formless clouds of neutral gas – the building blocks of later galaxies – were suddenly illuminated from within and heated by the energy of newborn suns. Eventually, this first generation of stars emitted so much high-energy ultraviolet radiation that matter in and around the protogalaxies was re-ionized. The hydrogen and helium atoms lost their electrons and the neutral gas turned into thin, hot plasma.

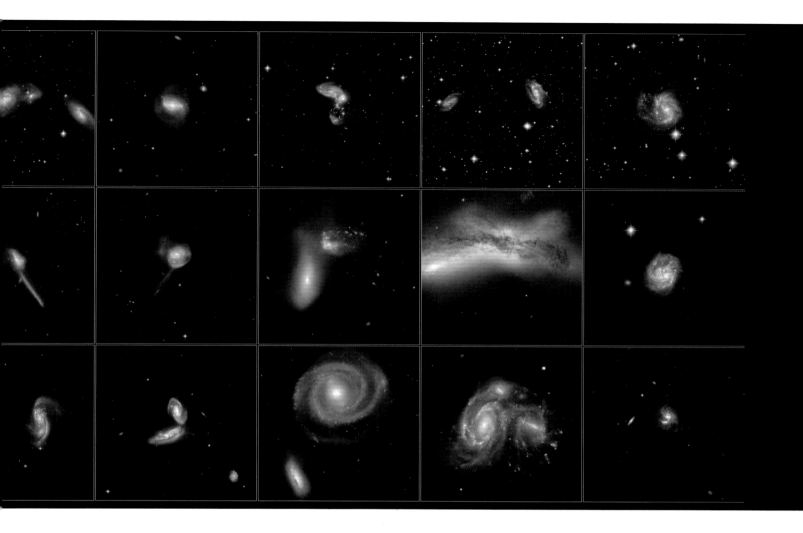

How exactly this re-ionization took place is one of the unsolved questions of cosmology. There were probably several hundred million years of expanding plasma bubbles in a sea of neutral gas that began to overlap over time. Research on the distribution of the neutral gas, both in space and in time, must explain the true connections for the exact timing and course of this process. However, this isn't easy. To look that far back in time, astronomers must observe signals that have been traveling for more than 13 billion years before reaching Earth. The small amount of radio radiation from the neutral gas decreased even further during this voyage. In addition, it was stretched in extremely long wavelengths. Such weak, low-frequency radio waves from the distant universe are easily masked by the radio radiation of the Milky Way and by terrestrial interference.

Special antennas, which can receive the longest waves and lowest frequencies, have been used for years to hunt down the much sought-after re-ionization signal. Scientists also discovered that the neutral gas, shortly before it was ionized again, leaves a kind of fingerprint in the spectrum of cosmic background radiation. These observations suggest that the age of re-ionization began about 180 million years after the Big Bang. It was also likely the period in which the very first stars in the universe began to shine. Incidentally, the measurements also proved that the neutral gas would have been colder than expected, possibly due to a special interaction with dark matter.

The future Australian Square Kilometre Array (SKA), which will be the largest radio observatory ever built, should be able to actually prove how this phase in the early evolution of the cosmos unfolded. Cosmologists hope to fully understand the process of galaxy formation – from the minimal density fluctuations in the newly born universe to the small, formless protogalaxies that appeared a few hundred million years after the Big Bang, seen by the Hubble Space Telescope and the ALMA Observatory. Even if we confirm how galaxies were formed and can understand how they have evolved over the course of cosmic history into their impressive diversity today, we'd always have this question: what does their future look like and what is the fate of the universe?

Dark Energy

The universe formed almost 14 billion years ago as a hot brew of hydrogen and helium atoms. Due to the expansion of the universe, this gas became thinner and thinner and cooled down. Influenced by the gravity of mysterious dark matter, it clumped together, becoming the first formless precursors of galaxies – the building blocks of the cosmos. In the course of many hundreds of millions of years, small protogalaxies merged into large, stately spirals like our Milky Way and gigantic elliptical galaxies like M87. In each of these galaxies, stars and planets were born out of the contracting gas clouds. There was life on at least one of these planets.

This is the history of the universe in a nutshell, as it has been discovered by astronomers during the last half century. It is an origin story in which gravity is the great master builder. The formation of the first protogalaxies, the collision and merger of galaxies, the formation of galaxy clusters and superclusters, the birth of stars and planets – everything is the result of gravity's ordering power and long reach. In this book, you can read about and admire the diversity of the cosmos and the richness of its structure – all shaped by gravity.

The evolution of the universe is still in progress. In the coming billions of years, galaxies will continue to collide, planets will clump together and supernovas will explode. But the largest birth wave of new stars is now billions of years behind us, and star formation activity in the universe will continue to decrease, from now into the distant future. That's not so surprising: at the end of its life, a star blows a portion of its matter back into space, while a portion remains locked in the mortal remains of the star – a white dwarf, a neutron star or a black hole. Therefore, over time, less and

A heavenly hermit

MCG+01-02-015, a few hundred million light-years from Earth, in the constellation Pisces, is a lonely object in a vast cosmic void, far from other galaxies. As a result of the accelerated expansion of the universe, all galaxies will live as loners in the distant future.

Cosmic expanses

The stars in this photo are part of our Milky Way; much further away (70 million light-years) is the galaxy NGC 1964. In the very distant future, other galaxies will no longer be visible from Earth, due to the accelerated expansion of the universe. This photo was taken with the European Southern Observatory's 7¾-foot (2.2 m) telescope at the La Silla Observatory in Chile.

Galactic driftwood

Unevenly shaped structures of gas and stars can be seen around the elliptical galaxy NGC 5291 in the southern constellation Centaurus. It is probably material that was thrown into space in the wake of a catastrophic cosmic collision hundreds of millions of years ago.

less material is available for the formation of new stars.

Galaxy collisions will also be less frequent in the distant future. Collisions often occur when large galaxy clusters first form, but ultimately most galaxies are devoured by the large elliptical galaxy in the heart of the cluster. Furthermore, the distances between galaxies are generally increasing as a result of the expansion of the universe. As space demands more and more space, the galaxies are pulled further apart and increasingly drift off as loners. Astronomers long thought that this isolation was only temporary, since all matter in the universe exerts gravity, however small it may be. The field equations of Albert

History with a tail

UGC 10214 is a galaxy 400 million light-years from Earth, in the constellation Draco. Because of its long veil of gas and stars, the galaxy has been given the nickname the "Tadpole Galaxy." The tail was created by the tidal effect of a small passing galaxy. In the distant future of the universe, such encounters will become increasingly rare.

Discontinuing business

This spiral galaxy in the constellation Hercules, situated about 300 million light-years from Earth, still shows considerable star-formation activity, as demonstrated by the bright blue star clusters in the spiral arms. In the future, however, all the existing gas will be used up and no new stars will be born in the galaxy. The bright star on the right belongs to our Milky Way.

Einstein's general theory of relativity show that gravity sets limits on the expansion of the universe. A hypothetically empty universe can in principle continue to grow at the same rate forever. However, once the cosmos is filled with matter, the rate of expansion must decrease over time. And if the average density of the universe is above a certain critical value, the expansion could even turn back into a contraction in the distant future.

According to this scenario, the galaxies will begin to converge again, and there will once again be collisions and mergers. This will result in new birth waves of stars. Over time, the universe will have shrunk so much that individual stars will collide. The cosmos will become a boiling sea of fire. When all the matter is pressed into a gigantic black hole, the film of the the Big Bang will run backward. Perhaps this "final bang" is the starting point for a new life cycle of the universe. To have a better view of the future evolution of the universe, astronomers must try to understand and explain the history of the universe's expansion. If they discover that the rate of expansion has fallen sharply in the past billions of years, this would indicate that the universe will shrink again sometime in the distant future. However, if they discover that the reduction in speed was much smaller in the past, this will indicate that the universe doesn't contain enough matter to really stop the expansion. In that case, there will never be a final bang.

The measurements are complicated. It took until the end of the last century for astronomers to obtain reliable results. With them, we gained a completely new picture of the evolution of the universe. Contrary to everyone's expectations, the speed of expansion didn't seem to decrease but continued to increase! For several billion years there has been an accelerated expansion. It seems that the braking effect of gravity has been overridden by a kind of anti-gravity in empty space – something Einstein had already pointed out.

The true nature of this mysterious dark energy is still unknown. Nor do we know whether there may be a relationship to the equally mysterious dark matter already mentioned in this book. The fact is that the universe hasn't become more comprehensible in recent decades. Dark energy and dark matter constitute the vast majority of the universe's content; the normal matter of stars, planets and living beings makes up only four percent of the total matter and energy content in the universe.

Precision measurements of cosmic background radiation also confirm the existence of large amounts of dark matter and dark energy. Only when these mysterious components are taken into account can the evolution and large-scale structure of the universe be properly understood. It's a sobering fact: cosmologists can describe the evolution of stars and galaxies in detail, but they only succeed in doing so if they include the effects of dark matter and dark energy in their calculations. However, no one actually understands the true nature of these enigmatic substances.

The discovery that the universe is constantly expanding at a faster rate is giving new light to the distant future and the fate of galaxies. It's a future full of emptiness, cold and darkness in which at some point the last star extinguishes, galaxies gradually decay and black holes slowly but surely evaporate. After an infinite time, the cosmos dies a silent death.

According to some cosmologists, it is possible that such an expanding and extinguishing universe could suddenly and unexpectedly produce a new Big Bang – a kind of reincarnation of the cosmos. Other scientists think that our universe is not unique, but rather part of an almost infinite multiverse of parallel universes. These theories are appealing to the imagination, but we have to question whether they'll ever be confirmed. The question of what happens in space or time outside our own universe may not even be suitable for scientific research, and we must leave it up to philosophy to look for an answer.

Be that as it may, it's a fascinating thought that our life is taking place in the early days of the universe: 14 billion years is an unimaginably long time by human standards, yet in a sense the force of the Big Bang hasn't yet subsided. Of course, Earth and life would never have come into being without a previous generation of stars producing elements such as carbon, oxygen and nitrogen through nuclear fusion processes. But we also owe our existence to the fact that supernovas still explode, stars are born and planets are formed in the Milky Way. In the very distant future of the universe this is no longer the case.

Here we sit on an inconspicuous crumb in an orbit around an inconspicuous star in a suburb of a completely average galaxy. *Homo sapiens* is nothing more than a small droplet in the vastness of an ocean, a cosmic ephemera that attempts to interpret a moment in the universe's record to unravel its past, present and future. Whether we will ever succeed is not known, but that's no reason to rest on our laurels.

"Remember to look up at the stars," Stephen Hawking once said, "and not down at your feet. Try to make sense of what you see and wonder about what makes the universe exist. Be curious."

A great view

Above the European Southern Observatory's Paranal Observatory in Chile rises the impressive and colorful ribbon of the Milky Way, the interior view of our own galaxy. Astronomers have succeeded in exploring the past, present and future of the universe in broad outlines from our subordinate place in space and time, particularly thanks to precise research on other galaxies.

INTERMISSION

Precision Cosmology

This hypnotic image comes from a computer simulation of the evolution of the universe. It shows how dark matter clumps to form a fibrous cosmic web. Galaxies are formed in the areas with the highest density (shown in green). Such simulations – combined with measurements of cosmic background radiation (the "afterglow" of the Big Bang), the distribution of galaxies in space and the expansion history of the universe – have resulted in today's standard model of cosmology. According to this model, only about five percent of the content of the universe is normal matter. The cosmos primarily contains large amounts of dark matter and dark energy. Although the true nature of these mysterious substances is still not unraveled, astronomers speak of precision cosmology.

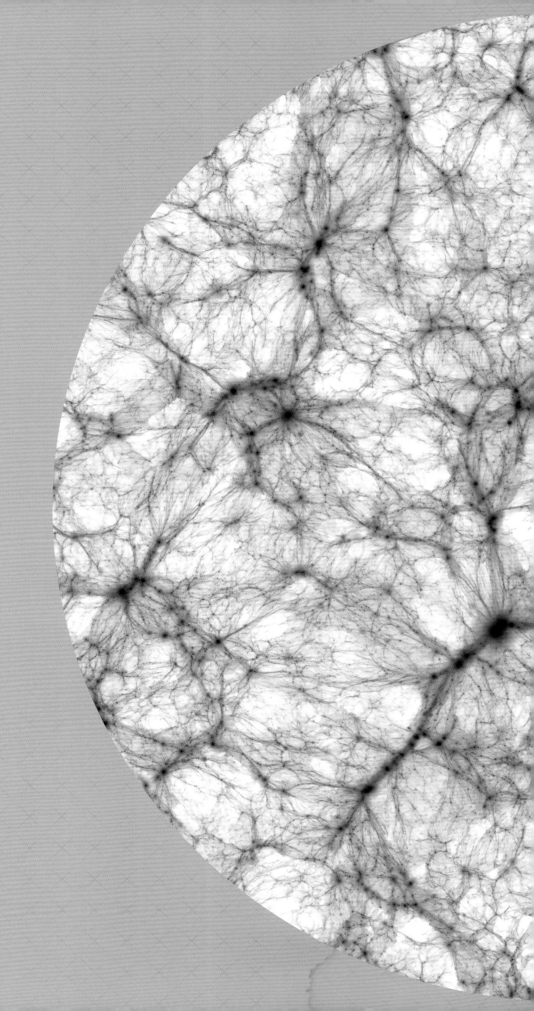

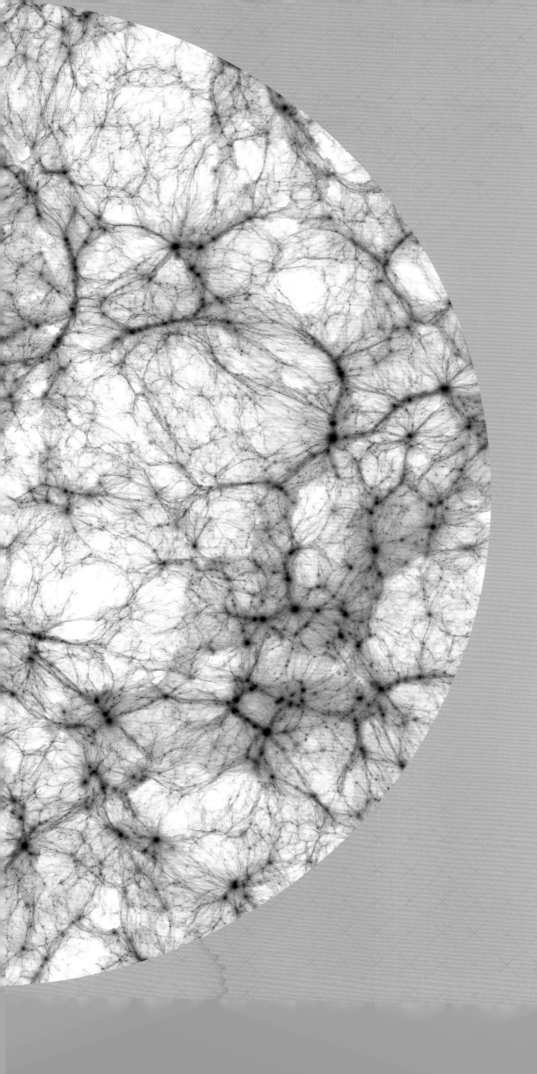

An angry giant

The gigantic elliptical galaxy NGC 5018 is surrounded by thin stripes and layers of star currents caused by tidal disturbances from neighboring galaxies. NGC 5018 is located 110 million light-years from Earth, in the constellation Virgo. The blue star in the upper right corner is a foreground star in our Milky Way. The image was taken with the European Southern Observatory's VLT Survey Telescope.

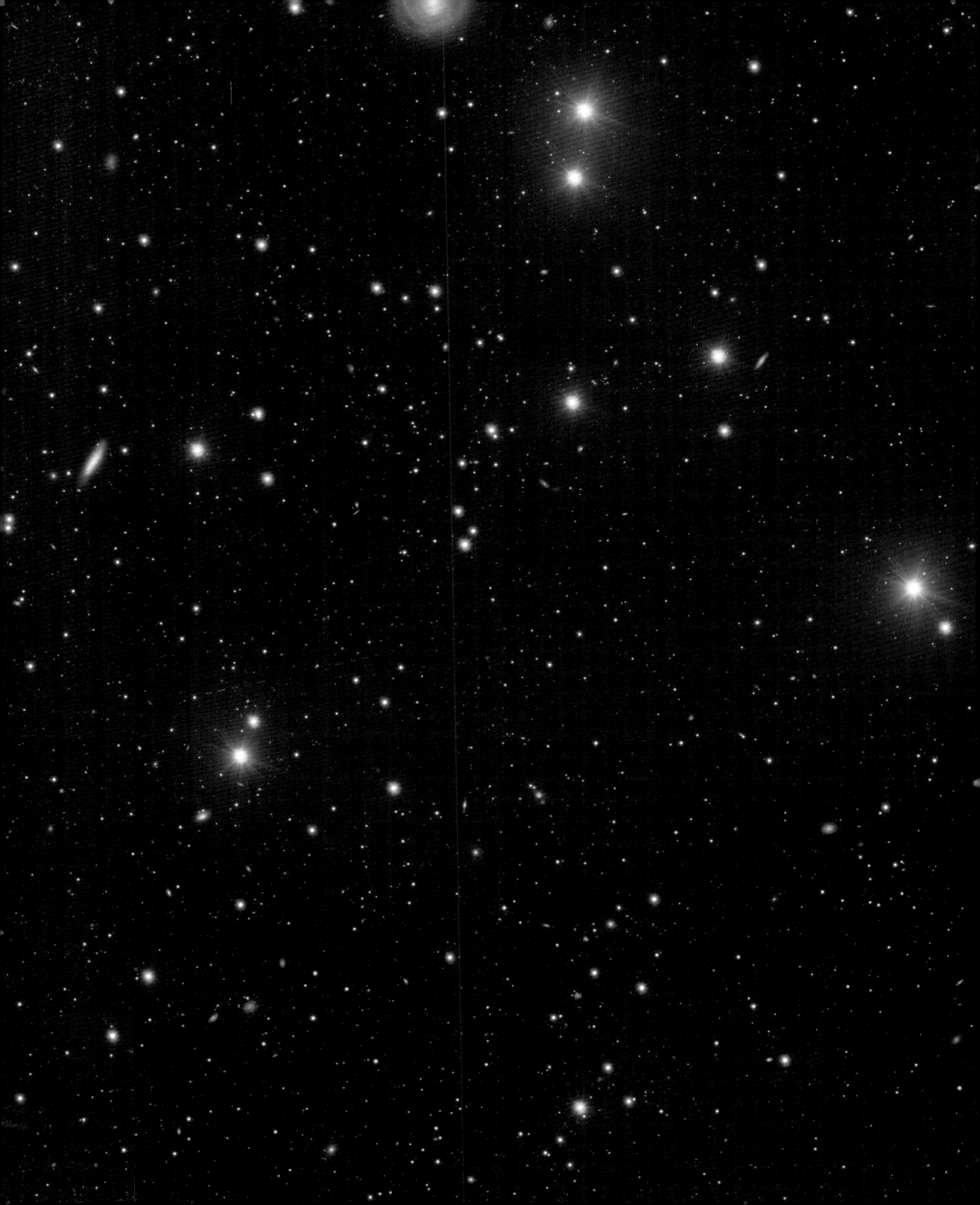

Picture Credits

p. 3: ESA/Hubble & NASA. p. 7: NASA/ESA/ M. Mutchler (STScI). p. 9: NASA/ESA/STScI. p. 10/11: ESO. p. 12/13: ESO. p. 14/15: ESO. p. 16/17: ESO/Y. Beletsky. p. 19: Rogelio Bernal Andreo. p. 20/21: NASA/ESA/N. Smith (University of California, Berkeley)/Hubble Heritage Team (STScI/AURA). p. 22: NASA/ESA/Orsola De Marco (Macquarie University). p. 23: NASA/ESA/Hubble Heritage Team. p. 24/25: ESA/Hubble/NASA. p. 26: ESA/Hubble/NASA/D. Padgett (GSFC)/T. Megeath (University of Toledo)/B. Reipurth (University of Hawaii). p. 27: NASA/JPL-Caltech. p. 28: Gemini Observatory/AURA/Lynette Cook. p. 29: ALMA (ESO/NAOJ/NRAO). p. 30: ESO/N. Bartmann. p. 32: NASA/ESA/Hubble Heritage Team (AURA/STScI). p. 33: ESA/Hubble/NASA/Gilles Chapdelaine. p. 34: NASA/JPL-Caltech/SSC/Judy Schmidt (Geckzilla). p. 35: T. A. Rector/University of Alaska, Anchorage/H. Schweiker/ NOAO/AURA/NSF. p. 36/37: NASA/ESA/G. Dubner (University of Buenos Aires) et al./A. Loll et al./T. Temim et al./F. Seward et al./NRAO/AUI/NSF/CXC/SSC/JPL-Caltech/XMM-Newton/ STScI. p. 38: NSF/Laser Interferometry Gravitational-wave Observatory/Sonoma State University/A. Simonnet. p. 39: NASA/GSFC/S. Wiessinger. p. 40/41: ESO/L. Calçada/M. Kornmesser. p. 43: NASA/JPL-Caltech. p. 44/45: ESO/S. Guisard (www.eso.org/~sguisard). p. 46: NASA/JPL-Caltech/ESA/CXC/STScI. p. 47: ESO/S. Gillessen et al. p. 48: ESO. p. 49: NASA/GSFC. p. 50/51: ESA/ATG medialab/ESO/S. Brunier. p. 52/53: ESA/Hubble/NASA. p. 54: V. Belokurov/D. Erkal (Cambridge, UK)/M. Putman (Columbia University, USA)/Axel Mellinger. p. 55: Atacama Large Millimeter/submillimeter Array (ALMA)/ESO/NAOJ/NRAO/B. Tafreshi (twanight.org). p. 56: Zdeňek Bardon/ESO. p. 57: ESO/R. Fosbury. p. 58/59: NASA/ESA/P. Crowther (University of Sheffield). p 60: NASA/ESA/A. Nota (STScI/ESA). p. 61: ESO. p. 62: NASA/JPL-Caltech/P. Barmby (CfA). p. 63: NAOJ/HSC Collaboration/Kavli Institute for the Physics and Mathematics of the Universe/STScI/Local Group Survey/NOAO/Digitized Sky Survey/Robert Gendler. p. 64/65: NASA/ESA/J. Dalcanton, B. F. Williams, L. C. Johnson (University of Washington, USA)/PHAT Team/R. Gendler. p. 66: NASA/ESA/Thomas M. Brown, Charles W. Bowers, Randy A. Kimble, Allen V. Sweigart (NASA GSFC)/Henry C. Ferguson (STScI).

p. 67: NASA/JPL-Caltech. p. 68: Johannes Schedler (Panther Observatory). p. 69: NASA/ESA/Z. Levay, R. van der Marel (STScI)/T. Hallas/A. Mellinger. p. 70: NASA/ESA/Hubble Heritage Team (AURA/STScI). p. 71: T. A. Rector (NRAO/AUI/NSF/NOAO/AURA)/M. Hanna (NOAO/AURA/NSF). p. 72/73: ESO. p. 75 top: ESA/Hubble/NASA. p. 75 bottom: NASA/JPL-Caltech/UCLA. p. 76/77: NASA/JPL-Caltech. p. 79: ESO/Digitized Sky Survey 2. p. 80/81: NASA/JPL-Caltech/UCLA. p. 82: NASA/JPL-Caltech/R. Hurt (SSC/Caltech). p. 83: NASA/ESA/A. Sarajedini (University of Florida)/Gilles Chapdelaine. p. 84: ESO/INAF-VST/OmegaCAM/A. Grado/L. Limatola/INAF-Capodimonte Observatory. p. 85: Virgo Consortium. p. 86/87: NASA/ESA/Hubble Heritage Team (STScI/AURA). p. 88/89: ESO. p. 91: ESA/Hubble/NASA/Judy Schmidt (Geckzilla). p. 92: NASA/ESA/A. Riess (STScI/JHU)/L. Macri (Texas A/M University)/Hubble Heritage Team (STScI/AURA). p. 93: NASA/ESA/Hubble Heritage Team (STScI/AURA)/Davide De Martin/Robert Gendler. p. 94/95: ESA/NASA. p. 96: NASA/ESA/Hubble Heritage Team (STScI/AURA)/M. Crockett, S. Kaviraj (Oxford University, UK)/R. O'Connell (University of Virginia)/B. Whitmore (STScI)/WFC3 Scientific Oversight Committee. p. 97: ESA/Hubble/NASA. p. 99: NASA/ESA/Hubble SM4 ERO Team. p. 100/101: NASA/ESA/Hubble Heritage Team (STScI/AURA). p. 102: ESA/Hubble/NASA/Judy Schmidt (Geckzilla). p. 103: ESO/Instrument Center for Danish Astrophysics/R. Gendler/J. E. Ovaldsen/C. Thöne/C. Feron. p. 104: NASA/ESA. p. 105: ESO. p. 106: NASA/ESA/Hubble Heritage Team (STScI/AURA). p. 107: NASA/ESA. p. 108: NASA/ESA/Andy Fabian (University of Cambridge, UK). p. 110/111: NASA/ESA/Hubble Heritage Team (STScI/AURA). p. 112: NASA/ESA. p. 113: ESA/Hubble/NASA. p. 115: ESO. p. 116: ESA/Hubble/NASA/LEGUS Team/R. Gendler. p. 117: NRAO/AUI/Erwin de Blok (ASTRON, Netherlands)/THINGS survey. p. 118/119: NASA/ESA/Hubble Heritage Team (STScI/AURA)/A. Zezas, J. Huchra (CfA). p. 120/121: NASA/ESA/Hubble Heritage Team (STScI/AURA)/William Blair (JHU). p. 122/123: ESO/Aniello Grado/Luca Limatola. p. 124/125: NASA/ESA/Hubble Heritage Team (STScI/AURA)/R. Gendler/J. GaBany. p. 126: NASA/ESA/Hubble Heritage Team (STScI/AURA). p. 127: NASA/ESA/Hubble Heritage Team

(STScI/AURA). p. 128/129: NASA/ESA/ S. Beckwith (STScI)/Hubble Heritage Team (STScI/AURA). p. 130: NASA/ESA/Hubble Heritage Team (STScI/AURA)/W. Keel (University of Alabama). p. 131: NASA/ESA/ Hubble SM4 ERO Team. p. 132: ESO. p. 134: Robert Gendler. p. 135: ESA/Hubble/NASA. p. 136: ESA/Hubble/NASA. p. 137: NASA/ESA/ A. Evans (Stony Brook University/University of Virginia/NRAO). p. 138/139: NASA/ESA/CXC/ JPL-Caltech. p. 140: NASA/ESA/Judy Schmidt (Geckzilla). p. 143: ESO. p. 144: ESA/Hubble/ NASA/Eedresha Sturdivant. p. 145: NASA/ESA/ Hubble Heritage Team (STScI/AURA)/P. Cote (Herzberg Institute of Astrophysics)/E. Baltz (Stanford University). p. 146/147: NASA/ESA/ S. Baum, C. O'Dea (RIT)/R. Perley, W. Cotton (NRAO/AUI/NSF)/Hubble Heritage Team (STScI/AURA). p. 148: ESA/Hubble/NASA. p. 149: NASA/ESA/Hubble Heritage Team (STScI/AURA)/R. O'Connell (University of Virginia)/WFC3 Scientific Oversight Committee. p. 151: NASA/ESA/M. Kornmesser. p. 152/153: NASA/ESA/M. Beasley (Instituto de Astrofísica de Canarias). p. 154: ESO/L. Calçada. p. 155: NASA/CXC/University of Wisconsin/ Y. Bai et al. p. 156: ESO/M. Kornmesser. p. 157: ESO/UKIRT Infrared Deep Sky Survey/SDSS. p. 158/159: ESO/L. Calçada. p. 160/161: ESO/ A. Grado/L. Limatola. p. 163: Rogelio Bernal Andreo. p. 164: NASA/ESA/Hubble Heritage Team (STScI/AURA)/K. Cook (Lawrence Livermore National Laboratory). p. 165: NASA/ ESA/Digitized Sky Survey 2/Davide De Martin. p. 166/167: ESO/INAF-VST/OmegaCAM/ Astro-WISE/Kapteyn Institute, University of Groningen. p. 168: Brent Tully/Daniel Pomarede. p. 169: NASA/CXO/Fabian et al./ Gendron-Marsolais et al./NRAO/AUI/NSF/ NASA/SDSS. p. 171: NASA/S. Habbal/ M. Druckmüller/P. Aniol. p. 172: ESA/Hubble/ NASA. p. 173: ESA/Hubble/NASA. p. 174: ESA/ J.-P. Kneib (Observatoire Midi-Pyrénées)/ Canada-France-Hawaii Telescope. p. 175: NASA/ESA/Johan Richard (Caltech)/Davide de Martin/James Long. p. 176: NASA/ESA/HST Frontier Fields Team (STScI). p. 177: NASA/ ESA/HST Frontier Fields Team (STScI). p. 178: NASA/ESA/Hubble Heritage Team (STScI/ AURA). p. 179: ESA/Hubble/NASA/HST Frontier Fields Team/Mathilde Jauzac (Durham University, UK/Astrophysics & Cosmology Research Unit, South Africa)/Jean-Paul Kneib (École Polytechnique Fédérale de Lausanne, Switzerland). p. 180: NASA/CXC/M. Markevitch et al./NASA/STScI/Magellan/University of Arizona/D. Clowe et al./ESO. p. 181: NASA/ ESA/R. Massey (Caltech). p. 182/183: NASA/ ESA/M. J. Jee, H. Ford (JHU). p. 184: NASA/ ESA/P. van Dokkum (Yale University). p. 185: NASA/ESA/CFHT/CXO/M. J. Jee (University of California, Davis)/A. Mahdavi (San Francisco State University). p. 186: CfA/V. de Lapparent et al. p. 187: Eagle Collaboration/Virgo Consortium. p. 188: Andrew Z. Colvin. p. 189: T. H. Jarrett (SSC). p. 190/191: Illustris Collaboration. p. 192: Matthew Colless/2dF/ Anglo-Australian Telescope. p. 193: TNG Collaboration. p. 194/195: ESA/Hubble/NASA/ D. Milisavljevic (Purdue University). p. 196/197: NASA/ESA/HST Frontier Fields Team (STScI)/ Judy Schmidt (Geckzilla). p. 198: STScI. p. 199: R. Williams (STScI)/Hubble Deep Field Team/ NASA/ESA. p. 200/201: NASA/ESA/S. Beckwith (STScI)/Hubble Ultra Deep Field Team. p. 202: NASA/ESA/J. Lotz (STScI). p. 203: ESA/Hubble/ NASA. p. 204/205: NASA/ESA/R. Windhorst, S. Cohen, M. Mechtley, M. Rutkowski (Arizona State University, Tempe)/R. O'Connell (University of Virginia)/P. McCarthy (Carnegie Observatories)/N. Hathi (University of California, Riverside)/R. Ryan (University of California, Davis)/H. Yan (Ohio State University)/A. Koekemoer (STScI). p. 207: ESA/ Hubble/NASA/A. Riess (STScI/JHU). p. 208: ESO/M. Kornmesser. p. 209: NASA/ESA and Hubble Heritage Team (STScI/AURA). p. 210/211: NASA/ESA/M. Kornmesser/ CANDELS Team (H. Ferguson). p. 212: NASA/ ESA/P. Oesch (Yale University). p. 213: Institute of Astronomy/Amanda Smith. p. 214: N. R. Fuller/NSF. p. 215: ESO/Joe DePasquale. p. 216/217: ESA/Planck Collaboration. p. 218: NRAO/AUI/NSF. p. 219: NASA/ESA/B. Salmon (STScI). p. 220/221: NASA/ESA/A. Evans (University of Virginia, Charlottesville/NRAO/ Stony Brook University)/Hubble Heritage Team (STScI/AURA). p. 222: ESA/Hubble/NASA/ N. Gorin (STScI)/Judy Schmidt (Geckzilla). p. 223: ESO/Jean-Christophe Lambry. p. 224: ESO. p. 225: NASA/Holland Ford (JHU)/ACS Science Team/ESA. p. 226/227: ESA/Hubble/ NASA/N. Grogin (STScI). p. 228: ESO/B. Tafreshi (twanight.org). p. 230/231: TNG Collaboration. p. 232/233: ESO/Marilena Spavone et al. p. 240: T.A. Rector (University of Alaska Anchorage)/H. Schweiker (WIYN/ NOAO/AURA/NSF).

Index

A

Abell, George 162, 165
Absolute zero (temperature) 220
Absorption lines 142
Accretion disk 47, 150
Active galaxy NGC 5128 (Centaurus A) 149
Active galaxies 149–150
Al-Sufi, Abd-al-Rahman 54, 62
Amino acids 31
Andromeda Galaxy, see Galaxies
Antennae galaxies 134–136
Anti-gravity, see Dark energy
Antiparticles 219
Arp, Hilton 136

B

Baade, Walter 180
Background radiation, cosmic 189, 193, 217, 219–221, 229–230
Barred spiral galaxies 56, 98–105, 210
 – M77 142
 – NGC 1015 206
 – NGC 1073 104
 – NGC 1300 100
 – NGC 1365 103
 – NGC 1398 105
 – NGC 4394 102
 – NGC 6217 98
Barnard, Edward 92
Bayer, Johannes 56
Bessel, Friedrich 114
Betelgeuse (star) 18
Big Bang 56, 189, 212, 214, 219–221, 229–230
Binary star 22, 31, 34, 41, 58
Black hole 8, 12, 31, 38, 41, 47, 48, 77, 124, 145–146, 149–157, 212, 222, 229; see also Sagittarius A*
 – Cygnus X-1 150
Blazar 149
Bode's Galaxy, see Spiral galaxies M81
Boötes void 169, 186
Bosma, Albert 117

C

Carbon monoxide 22
Carbon 37, 39
Carina Nebula 20
Cavities, cosmic (voids) 169, 186, 222
Cartwheel Galaxy (ESO 350-40) 136, 141
Centaurus A, see Galaxy NGC 5128
Centaurus Cluster 162, 165
Cepheid stars (variable stars) 61, 66, 86
Central thickening (of spiral galaxies) 93
Cigar Galaxy, see Galaxy M82
Cilicon 39
Clusters (of stars, galaxies) 6, 8, 161–193, 178, 222
Collisions (between galaxies) 134–142, 218, 220, 222, 224
Comet 18
Computer simulations 85, 136, 187, 192–193, 196, 198
Consciousness 31
Constellation
 – Adler 90
 – Andromeda 62
 – Aquarius 141
 – Aries 74
 – Boötes 157
 – Cancer 137
 – Canes Venatici 90, 124, 128, 133
 – Carina 20, 78, 181
 – Cassiopeia 62, 69, 80
 – Centaurus 82, 84, 144, 224
 – Cetus 142, 206
 – Chemical Furnace 78, 103, 105, 144, 160, 200, 204
 – Coma Berenices 7, 162
 – Corvus 134, 136
 – Cygnus 35, 144
 – Draco 78, 174–175, 224
 – Eridanus 88, 100
 – Gemini 62
 – Grus 202
 – Hercules 33, 166, 176, 226
 – Hydra 121
 – Leo 78, 93, 116
 – Octans 14
 – Orion 18, 58, 62
 – Pavo 10
 – Pegasus 130, 194
 – Pisces 183, 222
 – Sagittarius 42, 78
 – Scorpio 42
 – Sculptor 78, 114
 – Scutum 177, 196
 – Serpens 23
 – Taurus 62, 74
 – Triangulum 70
 – Ursa Major 62, 69, 90, 94, 96, 112, 118, 172, 197, 204
 – Ursa Minor 78
 – Virgo 92, 107, 109, 142, 144, 148, 162, 232
 – Volans 133
Contraction of the universe 229
Cosmic web 192, 230
Cosmos 6, 8, 31, 39, 106, 229, 230
Crab Nebula (M1) 37
Coma Cluster 162, 164–165, 180, 186, 188
Curtis, Heber 142

D

de Magalhães, Fernão 54
Deneb (star) 62
Denebola (star) 162
Deuterium (heavy hydrogen) 29
Dickinson, Mark 203
Dreyer, John 74, 142
Dark energy 222–231
Dark matter 85, 114–121, 178–185, 190, 192, 219–221, 229–231
Dwarf galaxy Holmberg II 112
Dwarf galaxy 78
Dwarf galaxies 78, 85, 98, 178
Dwarf planet Pluto 98, 206
Dwarf star Proxima Centauri 31, 141, 206
Dwingeloo 1 and 2, see Spiral galaxies

E

$E = mc^2$ 26, 219, 224
Eagle Nebula (M16) 22–23, 25
EAGLE simulation 187
Einstein, Albert 22, 26, 47, 150, 170, 173, 181, 214, 224, 229

Einstein ring 170, 172–173
Electromagnetic radiation 41
Elementary particles 219
Elliptical galaxies 102–113, 178, 210
– M60 107
– M87 (Virgo A) 142–143, 149–150, 154, 157, 162, 222
– NGC 1275 109, 169
– NGC 4696 113
– NGC 4874 165
– NGC 4889 165
– NGC 5018 238
– NGC 5291 224
– NGC 5866 106
Emission lines 142, 148
Eocene 194
Epsilon Virginis (star) 162
Eratosthenes 22
Earth 3, 33, 37, 39, 42, 90, 229
Eridanus Supervoid 188
Evolution of the universe 31, 33, 141, 196–231
Exoplanets 31
Expansion of the universe 122, 209, 214–215, 219–224, 229–230

F
Fingers-of-God 186, 188
Ford, Kent 117, 180
Fornax Cluster 160, 162, 178
Frontier Fields program 174, 176–177, 196, 202, 205

G
Galaxies 6, 8–9, 186, 212, 222
– A2744_YD4 208
– Andromeda Galaxy (M31, NGC 214) 62–69, 72, 74, 77–78, 85–86, 90, 117, 127, 150, 196, 206
– ESO 520-G13 126
– M32 66, 150
– M82 138
– M83 121
– M110 68
– MCG + 01-02-015 222
– Milkomeda 69
– NGC 1052-DF2 184
– NGC 205 68
– NGC 1232 88
– NGC 1277 152
– NGC 1316 178
– NGC 1964 223
– NGC 2467 22
– NGC 2623 137
– NGC 2841 96
– NGC 3314 130
– NGC 3344 97
– NGC 3621 25
– NGC 4038 134–136
– NGC 4039 134–136
– NGC 4298 7
– NGC 4302 7
– NGC 5128 (Centaurus A) 12
– NGC 5195 128, 133
– NGC 5584 92
– NGC 6611 25
– NGC 6744 10
– NGC 6814 90
– NGC 7098 14
– NGC 7252 141
– TON 618 154
– Triangulum Galaxy (M33, NGC 598) 70–78
– UGC 10214 225
Galaxy Clusters
– Abell 2218 174–175
– Abell 2744 (Pandora's Cluster) 179
– Abell 520 185
– Abell S1063 202
– MACS J0416.1 + 2403 179
– MACS J1149 + 2223 177
– MACS J717.5 + 3745 176
– ZwCl 0024 + 1652 183
Galaxy Redshift Survey (2dF) 192
Gamma radiation 49, 219
Gas nebula 6, 52, 66, 222
Gas cloud G2 48
Geller, Margaret 186, 188
Giant stars 37, 39, 47, 52, 75
– R136a1 58
– V838 Monocarotis 32
Global climate 31
Globular cluster (1E0657-558) 180–182
Globular cluster 12, 33, 82–83, 98, 146, 184
Gold 41
GOODS (Great Observatories Origins Deep Survey) 204
Gravitational field 150, 180
Gravitational lens 8, 170–177, 180–183, 202–205, 219
Gravitational waves 22, 25, 29, 31, 38, 41, 47, 114, 117, 130, 136, 169–170, 189, 222, 224
Gunn, James 183

H
Halo 8, 93, 106, 116
Hawking, Stephen 22, 229
Heavy metals 22, 41
Helium, Gas 22, 31, 37, 39, 56, 189, 214, 219–220
Helix Nebula 34, 37
Hercules Cluster 146, 162, 166
Herschel, William 35, 114, 134
Heart and Soul Nebula 80
Hipparchus 50
Homo sapiens 18, 33, 126, 229
Hubble Deep Field 199, 204
Hubble, Edwin 66, 86, 90, 98, 130, 142, 149, 210
Hubble sequence 99–102
Hubble Space Telescope, see Telescopes
Hydrogen, Gas 10, 22, 31, 37, 39, 54, 56, 189, 214, 219–220
Hydrogen bomb 29
Huchra, John 186, 188
Huygens, Christiaan 22
Hydra Cluster 162
Hydrocarbon 31

I
Infrared radiation 37, 42, 46–47, 77, 97
Interstellar matter 141
Intracluster gas 169, 178, 180, 189, 193
Ionization 214, 220–221
Isfahan 54

J
Jets 26, 31, 144, 145, 148, 150

K
Kapteyn, Jacobus 42

L

Lagoon Nebula 9
Laniakea Supercluster 168–169, 186
Lapparent, Valerie de 186, 188
Le Verrier, Urbain 114
Leavitt, Henrietta 61, 66
Life (in the universe) 25, 31, 37, 39, 229
Lemaître, Georges 214
Light arcs 172, 174–175, 180
Light echo 32
Light year 18
Light curvature 170, 172
Lenticular galaxies 106–113, 130
Local group (of galaxies) 77–78, 130, 165
Local superclusters 165
Lord Rosse 90, 133
Lynden-Bell, Donald 150

M

Maffei 1 and 2, see Spiral galaxies
Maffei, Paolo 78
Magellanic Clouds 52–61, 78, 133, 212
Magnetic field 22, 39, 109
MASS Redshift Survey 188
Massive star 39
Meathook galaxy 133
Mercury 37
Messier, Charles 74, 142, 162
Meteor 18
Milky Way 6, 8, 10, 16–26, 31, 33–34, 42–51, 77, 85, 90, 98, 104, 150, 168, 186, 212, 215, 222–223, 229
Milkomeda, see Galaxies
Molecular clouds 22
MOND (MOdified Newtonian Dynamics) theory 181, 183
Moon 133
Multiverse 229

N

Neptune 114
Neutron star 37–39, 41, 58, 150, 180, 222
New General Catalogue (NGC) 74
Newton, Isaac 181
Neyman, Jerzy 165
Nitrogen 37
Nova 62
Nuclear fusion 26, 29, 39, 220, 229

O

Observation horizon 203
Observatories, see Telescopes
Observatory, see Telescopes
Oort, Jan 42, 49, 114, 180, 184
Organic molecules 37, 41
Orion Nebula (M42) 18, 26–27, 58, 75
Oxygen 37, 39

P

Pandora's Cluster, see Galaxy Cluster Abell 2744
Particle accelerator 183
Perseus-Pisces Supercluster 152, 162, 165, 169, 188
Photonen 206, 219
Photosynthese 31
Pigafetta, Antonio 56
Planetary nebula 34–35, 37, 39, 178
Planets 26–35, 41
Plasma 219–221
Platinum 41
Pluto, see Dwarf planet
Precision cosmology 230
Primordial soup 212, 219
Protogalaxy, Protostars 23, 26–27, 29, 31, 41, 60, 205, 212, 219, 220–222
Proton 31
Proxima Centauri, see Dwarf star
Ptolemäus, Claudius 70, 74
Pulsar 37, 41

Q

Quasars 8, 148–150, 154, 172
– 3C273 148, 212
– ULAS J1120 + 0641 156–157

R

Radio astronomy 42, 148
Radio galaxies 144, 149
– Centaurus A 144
– Cygnus A 144
– Virgo A (M87), see Elliptical galaxy M87
Radio radiation 37, 39, 42, 144, 221
Radio telescopes 42, see also Telescopes
Redshift 186, 188, 206, 209, 212
Rees, Martin 150
Reionisation 210, 220–221
Religion 214
Rho Ophiuchi cloud complex 44
Rings 29, 67
Rotational energy 31, 98, 114, 116–117, 121
Rubin, Vera 117, 180

S

Sagittarius 78, 82, 83
Sagittarius A* (Black hole, Milky Way center) 42, 46–47, 49, 68, 150, 155
Sagittarius dwarf galaxy, see Dwarf galaxy
Satellite galaxies 78–87
Schmidt, Maarten 148, 169
Scott, Elizabeth 165
Soap Bubble Nebula 35, 37
Seyfert, Carl 142
Seyfert galaxies 124, 142, 149
– NGC 1097 144
Shapley, Harlow 78, 82
Sirius 114
Sloan Digital Sky Survey 82
Sloan Great Wall (Galaxienhaufen) 188
Sombrero Galaxy M104 110, 113
Solar eclipse 170
Solar system 31, 90
Space telescope, see Telescopes
Space-time, Curvature of 38, 41, 47, 170, 219–221
Spectroscopy 142, 165–166, 209
Spectrum (of stars) 103, 221
Speed of light 206
Spiral arms 10, 71, 88, 90, 92–93, 96, 124, 130, 133, 226
Spiral galaxies 6, 8, 14, 42, 74, 78, 90–113, 117, 127, 142, 178, 210, 226

- Arp 273 127
- Dwingeloo 1 and 2 78; M33 74, 77
- M66 93
- M81 118, 138
- M87, see Elliptical galaxy M87
- M96 116
- M101 94
- M106 124
- Maffei 1 and 2 78, 80
- NGC 2442 133
- NGC 4647 107
- NGC 4911 164
- NGC 7331 194

Spyromilia, Jason 18
Stephans Quintett 131, 133
Star S2 42
Star formation 22, 25–41, 52, 58, 72, 141, 205, 226
Star-forming regions
- N158 57
- N159 52, 57
- N160 57
- NGC 346 60
- NGC 604 70, 72–73, 75, 77

Star cluster 8, 22, 25, 47, 141, 178, 226
- 47 Tucanae 61
- M54 78, 82
- M92 33, 74
- NGC 2070 58
- Omega Centauri 82, 84
- Palomar 12 78, 82
- Pleiades (M45) 74
- Terzan 5 83
- Terzan 7 78, 82

Star streams 82, 85
Stellar wind 39, 75
Southern Pinwheel Galaxy, see Galaxy M83
Sun 18, 22, 25, 31, 33, 34, 37, 42, 90, 93, 103, 141, 170
Superclusters 6, 165, 168, 186
Supermassive black holes 150–157
Supernova 9, 37, 39, 41, 58, 75, 77, 92, 141, 150, 157, 174, 177–178, 222, 229
- 1987A 58

T

Tadpole Galaxy, see Galaxy UGC 10214

Tarantula Nebula 56–58, 77, 212
Telescopes, radio telescopes, space telescopes, observatories
- ALMA observatory, Chile 22, 29, 55, 208, 212–213, 218, 221
- Blanco Telescope, Chile 184
- Canada France Hawaii Telescope 174
- Cerro Paranal, Chile (VLT, ESO) 16, 18, 25, 142, 160, 166
- Chandra X-ray Observatory 75
- Dwingeloo telescope 78
- Euclid, Space Telescope 184
- European Southern Observatory (ESO) 8
- Event Horizon Telescope 49
- Extremely Large Telescope (ELT, ESO), Chile 158, 212, 215
- Fermi Space Telescope 49
- Gaia, Space Telescope 50
- Hale Telescope 90
- Hooker Telescope 66, 98
- Hubble Space Telescope 7–9, 58, 60–61, 64, 66, 68–69, 86, 90, 94, 98, 113, 133, 135, 148, 149, 172, 174–176, 181, 184, 196–200, 203–205, 209, 212, 220–221
- James Webb Space Telescope 158, 196, 205, 212
- Keck Telescope, Hawaii 204
- Kepler, Space Telescope 31
- La Silla Observatory (ESO), Chile 10, 223
- Large Synoptic Survey, Chile 184
- GALEX, Space Telescope 67
- Spitzer, Space Telescope 27, 42, 62
- Wise, Space Telescope 80
- Palomar-Observatorium 162
- Paranal-Observatorium, Chile 229
- Planck, Space Telescope 217
- Radcliffe Observatory 58
- Schmidt telescope 162
- SKA (Square Kilometer Array, Radio observatory) 221
- Very Large Telescope, Chile 184, 232
- Westerbork Radio Telescope 117

Theory of relativity 170
Tidal forces 56, 78, 82, 85, 127, 133–136, 224, 232
Three-dimensional image of the cosmos 181–189
Toomre, Alar & Juri 136
Trappist-1 31
Triangulum Galaxy, see Constellations and Galaxies
Tuning fork diagram from Hubble 98, 103, 210
Tyson, Anthony 180

U

Ultraviolet radiation 37, 52, 67, 77, 97, 214, 220
Universe 8–9, 18, 229
Universe, Structure of 6, 39, 49, 85, 90, 126, 186–193, 206, 219
Uranus 114

V

van Maanen, Adriaan 74
Vega (star) 62
Venus 37
Virgo A, see Elliptical galaxy M87
Virgo (Super)Cluster 107, 145, 162, 169

W

Whirlpool Galaxy
- M51 90, 128
- NGC 300 114

White dwarf 37, 103, 222
Wavelength, Change of 117, 122
Whirlpool Galaxy 90
Williams, Bob 204

X

X-rays 31, 37, 41–42, 46, 49, 75, 77, 149–150, 169
X-ray telescopes 180

Z

Zwicky, Fritz 114, 178, 184